U.S.A.
MATHEMATICAL OLYMPIADS,
1972–1986

NEW MATHEMATICAL LIBRARY

PUBLISHED BY

THE MATHEMATICAL ASSOCIATION OF AMERICA

The New Mathematical Library (NML) was started in 1961 by the School Mathematics Study Group to make available to high school students short expository books on various topics not usually covered in the high school syllabus. In a decade the NML matured into a steadily growing series of some twenty titles of interest not only to the originally intended audience, but to college students and teachers at all levels. Previously published by Random House and L. W. Singer, the NML became a publication series of the Mathematical Association of America (MAA) in 1975. Under the auspices of the MAA the NML continues to grow and remains dedicated to its original and expanded purposes. In its third decade, it contains some thirty titles.

U.S.A.
MATHEMATICAL OLYMPIADS,
1972–1986

Compiled and with solutions by

Murray S. Klamkin

University of Alberta

33

THE MATHEMATICAL ASSOCIATION
OF AMERICA

Problems about a triangle ABC on the surface of a sphere can often best be solved by considering the wedge shown in the cover illustration and the central angles α, β, and γ. This image is courtesy of James Hoffman, GANG, University of Massachusetts, Amherst, ©Geometry, Analysis, Numerics, and Graphics Group, UMASS, 1988.

Third Printing
©1988 by the Mathematical Association of America, Inc.
All rights reserved under International Pan-American Copyright Conventions.
Published in Washington, DC by
The Mathematical Association of America

Library of Congress Catalog Card Number 88-62611
Complete Set ISBN: 0-88385-600-X
Vol. 33 ISBN: 0-88385-634-4

Manufactured in the United States of America

Dedicated to the memory of

Samuel L. Greitzer

first chairman of the U.S.A. Mathematical Olympiad Committee
and long time friend and colleague.

NEW MATHEMATICAL LIBRARY

Other titles in preparation.

Editors' Note

The MAA is pleased to publish this collection of the first fifteen U.S.A. Mathematical Olympiads, prepared by the dedicated problem poser and solver, Murray S. Klamkin, who gave us NML vol. 29, the *International Mathematical Olympiads*, 1975–1985.

The Editors of this collection are grateful to Professor Klamkin for allowing them to indulge in their traditional role of occasionally elaborating on solutions and modifying problem statements.

Professor Klamkin, in his preface, encourages readers to let him know of improvements or generalizations they might think of to the solutions in this collection. The editors will be happy to include such additions, recommended by the author, in future reprintings of this volume.

The NML Editorial Committee is concerned that there are so few women contestants in the USAMO. We strongly recommend that the USAMO Committee examine the selection and recruitment policies in various schools and devise ways for attracting girls to problem solving.

We close this note by citing Professor G. Szegö's concluding remarks from his preface to NML volumes 11 and 12:

"We should not forget that the solution of any worthwhile problem very rarely comes to us easily and without hard work; it is rather the result of intellectual effort of days or weeks or months. Why should the young mind be willing to make this supreme effort? The explanation is probably the instinctive preference for certain values, that is, the attitude which rates intellectual effort and spiritual achievement higher than material advantage. Such a valuation can be only the result of a long cultural development of environment and public spirit which is difficult to accelerate by governmental aid or even by more intensive training in mathematics. The most effective means may consist of transmitting to the young mind the beauty of intellectual work and the feeling of satisfaction following a great and successful mental effort. The hope is justified that the present book might aid exactly in this respect and that it represents a good step in the right direction."

Joanne Elliott
Herbert Greenberg
Basil Gordon
Ivan Niven
Peter Ungar
Anneli Lax

October, 1988

Preface

This NML volume contains the first 15 U.S.A. Mathematical Olympiads (USAMO), 1972–1986 with solutions. For a history of various national mathematical olympiads as well as the International Mathematical Olympiads (IMO), see *CMI Report on the Mathematical Contests in Secondary Education I*, edited by Hans Freudenthal in Educational Studies in Mathematics 2 (1969) pp. 80–114, *A Historical Sketch of the Olympiads, National, and International*, by Nura D. Turner in Amer. Math. Monthly 85 (1978) pp. 802–807, and my Olympiad Corner #1–#80, in Crux Mathematicorum 1979–1986, in particular, #3, #4, #8, #58, #68 and #78 (in #8 there are also references to papers by S. L. Greitzer on the results of the first seven USAMO's as well as 102 references to mathematical competitions).

In the 1960's, Nura D. Turner was actively campaigning for starting a USAMO and for participating in the IMO. At that time there was negligible support for this by members of the National Contest Committee (NCC). In 1968, C. Salkind, the then chairman of the NCC, appointed a subcommittee consisting of N. Mendelsohn and N. D. Turner to explore the matter even though he was against starting these new activities. After the publication of *Why can't we have a USA Mathematical Olympiad?* by N. D. Turner in Amer. Math. Monthly 78 (1971) pp. 192–195, J. M. Earl, the then chairman of the NCC, reconstituted the Olympiad Subcommittee with S. L. Greitzer (chairman), A. Kalfus, H. Sagan and N. D. Turner. Being a member of the Board of Governors of the M.A.A., and unaware of these latter activities and not too pleased with the multiple choice National Mathematics Contest, I had written to Henry Alder, secretary of the M.A.A., to place a discussion of mathematical contests on the agenda of the next board meeting to be held in the summer of 1971 at Pennsylvania State University. Instead of contests being on the agenda, I ended up as a new member of the Olympiad Subcommittee as well as the NCC. At our first meeting in 1971, we voted, I believe 3–2, to start a USAMO. This was subsequently approved by the M.A.A.

The Olympiad itself was to consist of five essay-type problems requiring mathematical power of the contestants and to be done in 3 hours (later changed to $3\frac{1}{2}$ hours). The purpose of the Olympiad was to discover and encourage secondary school students with superior mathematical talent, students who possessed mathematical creativity and inventiveness as well as competence in mathematical techniques. Also, the top eight contestants could be selected as team members for subsequent participation in the IMO. Participation in the USAMO was to be by invitation only and limited to about 100 students selected from the top of the Honor Roll on the American High School Mathematics Examination (AHSME) plus possibly a few students of superior ability selected from those states that did not participate in the AHSME but had their own annual competitions. The eligibility requirements are now different due to the introduction in 1983 of an intermediate competition, the American Invitational Mathematics Examination (AIME) consisting of 15 problems of which only the numerical answer is to be determined. These problems are more challenging than those in the AHSME (a multiple choice exam) but not than those in the USAMO. All regularly enrolled secondary school students in the USA and Canada are eligible to write the AHSME, and the number who do so is on the order of 400,000. Those students who obtain at least a mark of 100 (out of 150) are invited to write the AIME, and the number who do so is on the order of 5,000. Those students who are then invited to write the USAMO are the 150 (approximately) who have the highest index score, the latter being the AHSME score plus 10 × the AIME score (max 15) so that each is equally weighted. The top 8 students in the USAMO are declared the winners and are invited to Washington, D.C. to receive prizes. The final wind up of the 3 competitions is an invited training session for 24 students (must be USA citizens or residents) to practise for the IMO competition. These consist of the top 8 students of the USAMO plus the few students, if any, who have obtained honorable mention and have participated in a previous IMO. The rest of the 24 places go to non-senior students at the top of the honor roll. If the 24 places have not been filled, then those non-senior students with the top combined index score of AHSME + 10 × AIME + 4 × USAMO (max 100) are selected. The 6 (previously 8) students who are to be the IMO team members are selected on the basis of the USAMO score plus the further tests at the training session.

For further information on these competitions as well as the American Junior High School Mathematics Examination (AJHSME), write to the Executive Director of the American Mathematics Competitions, Professor Walter E. Mientka, Department of Mathematics and Statistics, University of Nebraska, Lincoln, Nebraska 68588-0322.

The first USAMO Committee consisted of S. L. Greitzer, chairman, A. Kalfus, N. D. Turner and myself. Even though, according to my files, S. L.

Greitzer was initially doubtful about starting a USAMO, once it was approved, he became a very efficient one-man administrator of it for many years subsequently. I had set the problems and they were checked out by A. Schwartz and D. J. Newman. Invitations were sent to 106 students on April 14, 1972, and 100 students took the First USA Olympiad on May 9, 1972. The papers were graded by J. Bender, A. Bumby, S. Leader, B. Muckenhoupt, and H. Zimmerberg of Rutgers University. The top papers were then regraded by L. M. Kelly of Michigan State University and myself.

The top eight contestants (their names and school affiliations as well as those of subsequent winners are listed in the Appendix) were honored in June 1972 in Washington D.C. at a prestigious three-part Awards Ceremony arranged very effectively by N. D. Turner (and for many years subsequently): the bestowing of awards and the giving of the Olympiad Address by Emmanuel R. Piore in the Board Room of the National Academy of Sciences, and the reception and dinner in the rooms of the Diplomatic Reception Area of the Department of State. The costs of these ceremonies were defrayed by a generous grant from the IBM Corporation. These ceremonies were repeated in successive years (with several exceptions when the reception and dinner were held in the Great Hall of the National Academy of Science) and defrayed by continued generous grants from IBM and MAA support. Some of the successive addresses were given by Saunders MacLane, Lowell J. Paige, Peter D. Lax, Andrew M. Gleason, Alan J. Hoffman, Ivan Niven, Charles Fefferman, Ronald L. Graham, and Neil J. Sloane.

During 1973–1985, the problems in the USAMO were set by an Examination Subcommittee consisting of three members (increased to four after 1982) of which I was chairman. Peter Paige, Cecil C. Rousseau, Tom Griffiths, Andy Liu, and Joseph Konhauser have served on this subcommittee during this 12 year period; I am pleased to acknowledge their long time collaboration. After my resignation from the committee, Ian Richards became the new chairman.

The solutions of the USAMO problems have been compiled annually from the solutions of the Examination Committee and made available in pamphlet form for a nominal charge. This compilation was first done by S. L. Greitzer up through 1982, by A. Liu and myself during 1983–5, and by C. C. Rousseau in 1986. Naturally, many of the solutions in this book are similar to those previously published. Moreover, the solutions have been edited, quite a number have been changed and/or extended, and references have been added where pertinent. Although the problems are arranged in chronological order at the beginning, the solutions are arranged by subject matter to facilitate the learning in a particular field. The subject matter classifications are Algebra, Number Theory, Plane Geometry, Solid Geometry, Geometric Inequalities, Inequalities, and Combinatorics & Probability. A previous example of this type of arrangement occurs in the recommended book [48]. Also, at the end is a Glossary of some frequently used terms and theorems as well as a comprehensive bibliography with items numbered and referred to in brackets in the text.

The solutions given here are more detailed than need be for the USAMO contestants, since it was the consensus of the NML Committee that this would make them more accessible and of service to a wider audience.

A frequent concern of contestants is how detailed a solution has to be to obtain full marks. This of course depends on the graders. One should avoid "hand waving" arguments and when in doubt, one should include details rather then leave them out. It is in this aspect that Olympiad type competitions are vastly superior to the AJHSME, AHSME, and AIME type competitions. In the former type, one has to give a *well written complete sustained argument* to obtain full marks whereas in the AJHSME and the AHSME one can just guess at the correct multiple choice answers, and in the AIME one does not have to substantiate any of the numerical answers. This ability to write is very important and is only recently being emphasized. I and many others do not care for multiple choice type competitions or even those like the AIME. Apparently, the raison d'être for these types of competitions is the relative ease of administering them to the very large number of contestants. This could be achieved by having run off competitions graded locally and only the final competition with a small number of contestants graded centrally. In the European socialist countries, where there is a strong tradition in mathematical competitions, there are separate Olympiads for the 7th, 8th, 9th, and 10th grades. It is very unfortunate that we do not have at the very least a junior Olympiad type competition as a follow up for the AJHSME.

In the grading of the USAMO there are extra marks for elegant solutions and/or non-trivial generalizations with proof. Although generalizations are part and parcel of mathematical creativity and elegant solutions are much more satisfactory and transparent than non-elegant ones, finding them usually takes extra time, unless one has some special a priori knowledge, or else one has had long practise in finding them. So contestants should not "go too far out of their way" looking for elegance or generalizations. Nevertheless, if there is time, contestants are advised to strive for refinements, since elegance is frequently a sign of real understanding and generalization a sign of creativity.

No doubt, many of the solutions given here can be improved upon or generalized, particularly with special a priori knowledge; when a reader finds this to be the case, I would be very grateful to receive any such communications.

I am greatly indebted to the late Samuel L. Greitzer and to Andy Liu for sharing the joys and burdens of the USAMO Committee as well as the joys and burdens of coaching the USA Olympiad teams and participating in the IMO from 1975–1980 and 1981–1984, respectively, and also to Walter E. Mientka, Executive Director of the American Mathematics Competitions, for his continued cheerful efficient cooperation over many years. I am also grateful to the Examination Committee members mentioned previously, Samuel L. Greitzer, Peter D. Lax, Andy Liu, Peter Ungar and all members of the NML Committee, particularly Anneli Lax and Ivan Niven for improvements and additions. Lastly but not least, I am very grateful to my wife Irene for her assistance with this book in many ways.

Murray S. Klamkin
University of Alberta

Contents

Problems

First U.S.A. Mathematical Olympiad, May 9, 1972

1972/1. (N.T./1)†. The symbols $(a, b, \ldots, g)$ and $[a, b, \ldots, g]$ denote the greatest common divisor and least common multiple, respectively, of the positive integers $a, b, \ldots, g$. For example, $(3, 6, 18) = 3$ and $[6, 15] = 30$. Prove that

$$\frac{[a, b, c]^2}{[a, b][b, c][c, a]} = \frac{(a, b, c)^2}{(a, b)(b, c)(c, a)}.$$

1972/2. (S.G./2). A given tetrahedron $ABCD$ is isosceles, that is, $AB = CD$, $AC = BD$, $AD = BC$. Show that the faces of the tetrahedron are acute-angled triangles.

1972/3. (C. & P./14). A random number selector can only select one of the nine integers $1, 2, \ldots, 9$, and it makes these selections with equal probability. Determine the probability that after n selections ($n > 1$), the product of the n numbers selected will be divisible by 10.

1972/4. (I/6). Let R denote a non-negative rational number. Determine a fixed set of integers a, b, c, d, e, f, such that for *every* choice of R,

$$\left| \frac{aR^2 + bR + c}{dR^2 + eR + f} - \sqrt[3]{2} \right| < \left| R - \sqrt[3]{2} \right|.$$

1972/5. (P.G./6). A given convex pentagon $ABCDE$ has the property that the area of each of the five triangles ABC, BCD, CDE, DEA, and EAB is unity. Show that all pentagons with the above property have

---

†N.T./1 refers to Number Theory, Problem 1; its solution is listed in the N.T. solution section, see table of contents.

the same area, and calculate that area. Show, furthermore, that there are infinitely many non-congruent pentagons having the above area property.

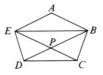

Second U.S.A. Mathematical Olympiad, May 1, 1973

1973/1. (G.I./7). Two points P and Q lie in the interior of a regular tetrahedron $ABCD$. Prove that angle $PAQ < 60°$.

1973/2. (A/10). Let $\{X_n\}$ and $\{Y_n\}$ denote two sequences of integers defined as follows:

$$X_0 = 1, \quad X_1 = 1, \quad X_{n+1} = X_n + 2X_{n-1} \quad (n = 1, 2, 3, \ldots),$$
$$Y_0 = 1, \quad Y_1 = 7, \quad Y_{n+1} = 2Y_n + 3Y_{n-1} \quad (n = 1, 2, 3, \ldots).$$

Thus, the first few terms of the sequences are:

$$X: \quad 1, 1, \ 3, \ 5, \ 11, \ 21, \ldots,$$
$$Y: \quad 1, 7, 17, 55, 161, 487, \ldots.$$

Prove that, except for the "1", there is no term which occurs in both sequences.

1973/3. (C. & P./12). Three distinct vertices are chosen at random from the vertices of a given regular polygon of $(2n + 1)$ sides. If all such choices are equally likely, what is the probability that the center of the given polygon lies in the interior of the triangle determined by the three chosen random points?

1973/4. (A/5). Determine all the roots, real or complex, of the system of simultaneous equations

$$x + y + z = 3,$$
$$x^2 + y^2 + z^2 = 3,$$
$$x^3 + y^3 + z^3 = 3.$$

1973/5. (N.T./6). Show that the cube roots of three distinct prime numbers cannot be three terms (not necessarily consecutive) of an arithmetic progression.

Third U.S.A. Mathematical Olympiad, May 7, 1974

1974/1. (A/7). Let a, b, and c denote three distinct integers, and let P denote a polynomial having all integral coefficients. Show that it is impossible that $P(a) = b$, $P(b) = c$, and $P(c) = a$.

1974/2. (I/1). Prove that if a, b, and c are positive real numbers, then

$$a^a b^b c^c \geqslant (abc)^{(a+b+c)/3}.$$

1974/3. (G.I./6). Two boundary points of a ball of radius 1 are joined by a curve contained in the ball and having length less than 2. Prove that the curve is contained entirely within some hemisphere of the given ball.

1974/4. (I/8). A father, mother and son hold a family tournament, playing a two person board game with no ties. The tournament rules are:

(i) The weakest player chooses the first two contestants.

(ii) The winner of any game plays the next game against the person left out.

(iii) The first person to win two games wins the tournament.

The father is the weakest player, the son the strongest, and it is assumed that any player's probability of winning an individual game from another player does not change during the tournament.

Prove that the father's optimal strategy for winning the tournament is to play the first game with his wife.

1974/5. (P.G./7). Consider the two triangles $\triangle ABC$ and $\triangle PQR$ shown in Figure 1. In $\triangle ABC$, $\angle ADB = \angle BDC = \angle CDA = 120°$. Prove that $x = u + v + w$.

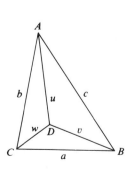
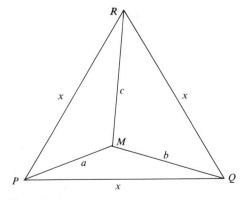

Figure 1

Fourth U.S.A. Mathematical Olympiad, May 6, 1975

1975/1. (N.T./9).   (a) Prove that

$$[5x] + [5y] \geq [3x + y] + [3y + x],$$

where $x, y \geq 0$ and $[u]$ denotes the greatest integer $\leq u$ (e.g., $[\sqrt{2}] = 1$).
 (b) Using (a) or otherwise, prove that

$$\frac{(5m)!(5n)!}{m!n!(3m + n)!(3n + m)!}$$

is integral for all positive integral m and n.

1975/2. (G.I./4).   Let A, B, C, D denote four points in space and AB the distance between A and B, and so on. Show that

$$AC^2 + BD^2 + AD^2 + BC^2 \geq AB^2 + CD^2.$$

1975/3. (A/8).   If $P(x)$ denotes a polynomial of degree n such that $P(k) = k/(k + 1)$ for $k = 0, 1, 2, \ldots, n$, determine $P(n + 1)$.

1975/4. (G.I./2).   Two given circles intersect in two points P and Q. Show how to construct a segment AB passing through P and terminating on the two circles such that $AP \cdot PB$ is a maximum.

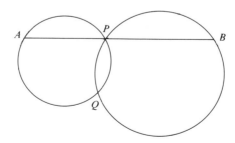

1975/5. (C. & P./13).   A deck of n playing cards, which contains three aces, is shuffled at random (it is assumed that all possible card distributions are equally likely). The cards are then turned up one by one from the top until the second ace appears. Prove that the expected (average) number of cards to be turned up is $(n + 1)/2$.

Fifth U.S.A. Mathematical Olympiad, May 4, 1976

1976/1. (C. & P./1).

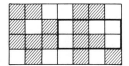

(a) Suppose that each square of a 4 × 7 chessboard, as shown above, is colored either black or white. Prove that with *any* such coloring, the board must contain a rectangle (formed by the horizontal and vertical lines of the board such as the one outlined in the figure) whose four distinct unit corner squares are all of the same color.

(b) Exhibit a black-white coloring of a 4 × 6 board in which the four corner squares of every rectangle, as described above, are not all of the same color.

1976/2. (P.G./4). If A and B are fixed points on a given circle and XY is a variable diameter of the same circle, determine the locus of the point of intersection of lines AX and BY. You may assume that AB is not a diameter.

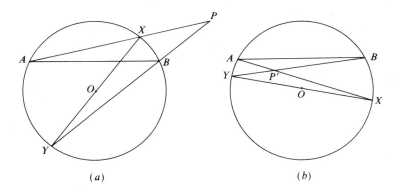

(a) (b)

1976/3. (N.T./2). Determine all integral solutions of

$$a^2 + b^2 + c^2 = a^2 b^2.$$

1976/4. (G.I./5). If the sum of the lengths of the six edges of a trirectangular tetrahedron $PABC$ (i.e., $\angle APB = \angle BPC = \angle CPA = 90°$) is S, determine its maximum volume.

1976/5. (A/9). If $P(x)$, $Q(x)$, $R(x)$, and $S(x)$ are all polynomials such that
$$P(x^5) + xQ(x^5) + x^2R(x^5) = (x^4 + x^3 + x^2 + x + 1)S(x),$$
prove that $x - 1$ is a factor of $P(x)$.

Sixth U.S.A. Mathematical Olympiad, May 3, 1977

1977/1. (A/6). Determine all pairs of positive integers (m, n) such that
$$(1 + x^n + x^{2n} + \cdots + x^{mn}) \text{ is divisible by } (1 + x + x^2 + \cdots + x^m).$$

1977/2. (P.G./3). ABC and $A'B'C'$ are two triangles in the same plane such that the lines AA', BB', CC' are mutually parallel. Let $[ABC]$ denote the area of triangle ABC with an appropriate $\pm$ sign, etc.; prove that
$$3([ABC] + [A'B'C']) = [AB'C'] + [BC'A'] + [CA'B']$$
$$+ [A'BC] + [B'CA] + [C'AB].$$

1977/3. (A/4). If a and b are two of the roots of $x^4 + x^3 - 1 = 0$, prove that ab is a root of $x^6 + x^4 + x^3 - x^2 - 1 = 0$.

1977/4. (S.G./3). Prove that if the opposite sides of a skew (non-planar) quadrilateral are congruent, then the line joining the midpoints of the two diagonals is perpendicular to these diagonals, and conversely, if the line joining the midpoints of the two diagonals of a skew quadrilateral is perpendicular to these diagonals, then the opposite sides of the quadrilateral are congruent.

1977/5. (I/4). If a, b, c, d, e are positive numbers bounded by p and q, i.e., if they lie in $[p, q]$, $0 < p$, prove that
$$(a + b + c + d + e)\left(\frac{1}{a} + \frac{1}{b} + \frac{1}{c} + \frac{1}{d} + \frac{1}{e}\right) \leqslant 25 + 6\left(\sqrt{\frac{p}{q}} - \sqrt{\frac{q}{p}}\right)^2$$
and determine when there is equality.

Seventh U.S.A. Mathematical Olympiad, May 2, 1978

1978/1. (I/2). Given that a, b, c, d, e are real numbers such that
$$a + b + c + d + e = 8,$$
$$a^2 + b^2 + c^2 + d^2 + e^2 = 16.$$
Determine the maximum value of e.

1978/2. (P.G./2). *ABCD* and *A'B'C'D'* are square maps of the same region, drawn to different scales and superimposed as shown in the figure. Prove that there is only one point *O* on the small map which lies directly over point *O'* of the large map such that *O* and *O'* each represent the same place of the country. Also, give a Euclidean construction (straight edge and compasses) for *O*.

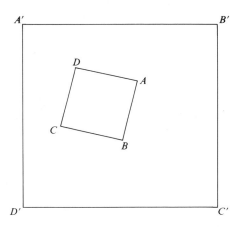

1978/3. (N.T./5). An integer *n* will be called *good* if we can write

$$n = a_1 + a_2 + \cdots + a_k,$$

where $a_1, a_2, \ldots, a_k$ are positive integers (not necessarily distinct) satisfying

$$\frac{1}{a_1} + \frac{1}{a_2} + \cdots + \frac{1}{a_k} = 1.$$

Given the information that the integers 33 through 73 are good, prove that every integer ≥ 33 is good.

1978/4. (S.G./6). (a) Prove that if the six dihedral angles (i.e., angles between pairs of faces) of a given tetrahedron are congruent, then the tetrahedron is regular.

(b) Is a tetrahedron necessarily regular if five dihedral angles are congruent?

1979/5. (C. & P./2). Nine mathematicians meet at an international conference and discover that among any three of them, at least two speak a common language. If each of the mathematicians can speak at most three languages, prove that there are at least three of the mathematicians who can speak the same language.

Eighth U.S.A. Mathematical Olympiad, May 1, 1979

1979 / 1. (N.T. / 3). Determine all non-negative integral solutions $(n_1, n_2, \ldots, n_{14})$ if any, apart from permutations, of the Diophantine equation

$$n_1^4 + n_2^4 + \cdots + n_{14}^4 = 1{,}599.$$

1979 / 2. (S.G. / 1). Let S be a great circle with pole P. On any great circle through P, two points A and B are chosen equidistant from P. For any *spherical triangle ABC* (the sides are great circle arcs), where C is on S, prove that the great circle arc CP is the angle bisector of angle C.

Note. A great circle on a sphere is one whose center is the center of the sphere. A pole of the great circle S is a point P on the sphere such that the diameter through P is perpendicular to the plane of S.

1979 / 3. (I / 9). Given three identical n-faced dice whose corresponding faces are identically numbered with arbitrary integers. Prove that if they are tossed at random, the probability that the sum of the bottom three face numbers is divisible by three is greater than or equal to $1/4$.

1979 / 4. (G.I. / 1). Show how to construct a chord BPC of a given angle A through a given point P such that $1/BP + 1/PC$ is a maximum.

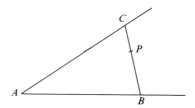

1979 / 5. (C. & P. / 6). A certain organization has n members, and it has $n + 1$ three-member committees, no two of which have identical membership. Prove that there are two committees which share exactly one member.

Ninth U.S.A. Mathematical Olympiad, May 6, 1980

1980/1. (A/1). A two-pan balance is inaccurate since its balance arms are of different lengths and its pans are of different weights. Three objects of different weights A, B, and C are each weighed separately. When placed on the left-hand pan, they are balanced by weights A_1, B_1, C_1, respectively. When A and B are placed on the right-hand pan, they are balanced by A_2 and B_2, respectively. Determine the true weight of C in terms of A_1, B_1, C_1, A_2 and B_2.

1980/2. (I/5). Determine the maximum number of different three-term arithmetic progressions which can be chosen from a sequence of n real numbers

$$a_1 < a_2 < \cdots < a_n.$$

1980/3. (A/11). Let $F_r = x^r \sin rA + y^r \sin rB + z^r \sin rC$, where x, y, z, A, B, C are real and $A + B + C$ is an integral multiple of π. Prove that if $F_1 = F_2 = 0$, then $F_r = 0$ for all positive integral r.

1980/4. (S.G./7). The inscribed sphere of a given tetrahedron touches all four faces of the tetrahedron at their respective centroids. Prove that the tetrahedron is regular.

1980/5. (I/3). Prove that, for numbers a, b, c in the interval $[0, 1]$,

$$\frac{a}{b+c+1} + \frac{b}{(c+a+1)} + \frac{c}{(a+b+1)} + (1-a)(1-b)(1-c) \leqslant 1.$$

Tenth U.S.A. Mathematical Olympiad, May 5, 1981

1981/1. (P.G./1). The measure of a given angle is $180°/n$ where n is a positive integer not divisible by 3. Prove that the angle can be trisected by Euclidean means (straight edge and compasses).

1981/2. (C. & P./7). Every pair of communities in a county are linked directly by exactly one mode of transportation: bus, train or airplane. All three modes of transportation are used in the county with no community being serviced by all three modes and no three communities being linked pairwise by the same mode. Determine the maximum number of communities in this county.

1981/3. (G.I./3). If A, B, C are the angles of a triangle, prove that

$$-2 \leqslant \sin 3A + \sin 3B + \sin 3C \leqslant 3\sqrt{3}/2,$$

and determine when equality holds.

1981/4. (S.G./5). The sum of the measures of all the face angles of a given convex polyhedral angle is equal to the sum of the measures of all its dihedral angles. Prove that the polyhedral angle is a trihedral angle.

Note. A convex polyhedral angle may be formed by drawing rays from an exterior point to all points of a convex polygon.

1981/5. (I/10). If x is a positive real number, and n is a positive integer, prove that

$$[nx] > \frac{[x]}{1} + \frac{[2x]}{2} + \frac{[3x]}{3} + \cdots + \frac{[nx]}{n},$$

where $[t]$ denotes the greatest integer less than or equal to t. For example, $[\pi] = 3$ and $[\sqrt{2}] = 1$.

Eleventh U.S.A. Mathematical Olympiad, May 4, 1982

1982/1. (C. & P./3). In a party with 1982 persons, among any group of four there is at least one person who knows each of the other three. What is the minimum number of people in the party who know everyone else.

1982/2. (A/12). Let $S_r = x^r + y^r + z^r$ with x, y, z real. It is known that if $S_1 = 0$,

$$(*) \qquad \frac{S_{m+n}}{m+n} = \frac{S_m}{m} \frac{S_n}{n}$$

for $(m, n) = (2, 3)$, $(3, 2)$, $(2, 5)$, or $(5, 2)$. Determine *all* other pairs of integers (m, n) if any, so that $(*)$ holds for all real numbers x, y, z such that $x + y + z = 0$.

1982/3. (G.I./10). If a point A_1 is in the interior of an equilateral triangle ABC and point A_2 is in the interior of $\triangle A_1 BC$, prove that

$$\text{I.Q.} (A_1 BC) > \text{I.Q.} (A_2 BC),$$

where the *isoperimetric quotient* of a figure F is defined by

$$\text{I.Q.}(F) = \text{Area}(F) / [\text{Perimeter}(F)]^2.$$

1982/4. (N.T./10). Prove that there exists a positive integer k such that $k \cdot 2^n + 1$ is composite for every positive integer n.

1982/5. (S.G./8). A, B, and C are three interior points of a sphere S such that AB and AC are perpendicular to the diameter of S through A, and so that two spheres can be constructed through A, B, and C which are both tangent to S. Prove that the sum of their radii is equal to the radius of S.

Twelfth U.S.A. Mathematical Olympiad, May 3, 1983

1983/1. (C. & P./11). On a given circle, six points A, B, C, D, E, and F are chosen at random, independently and uniformly with respect to arc length. Determine the probability that the two triangles ABC and DEF are disjoint, i.e., have no common points.

1983/2. (A/2). Prove that the roots of

$$x^5 + ax^4 + bx^3 + cx^2 + dx + e = 0$$

cannot *all* be real if $2a^2 < 5b$.

1983/3. (C. & P./8). Each set of a finite family of subsets of a line is a union of two closed intervals. Moreover, any three of the sets of the family have a point in common. Prove that there is a point which is common to at least half the sets of the family.

1983/4. (S.G./4). Six segments S_1, S_2, S_3, S_4, S_5 and S_6 are given in a plane. These are congruent to the edges AB, AC, AD, BC, BD and CD, respectively, of a tetrahedron $ABCD$. Show how to construct a segment congruent to the altitude of the tetrahedron from vertex A with straight-edge and compasses.

1983/5. (N.T./11). Consider an open interval of length $1/n$ on the real number line, where n is a positive integer. Prove that the number of irreducible fractions p/q, with $1 \leqslant q \leqslant n$, contained in the given interval is at most $(n + 1)/2$.

Thirteenth U.S.A. Mathematical Olympiad, May 1, 1984

1984/1. (A/3). The product of two of the four roots of the quartic equation $x^4 - 18x^3 + kx^2 + 200x - 1984 = 0$ is -32. Determine the value of k.

1984/2. (N.T./7). The geometric mean of any set of m non-negative numbers is the m-th root of their product.

(i) For which positive integers n is there a finite set S_n of n distinct positive integers such that the geometric mean of any subset of S_n is an integer?

(ii) Is there an infinite set S of distinct positive integers such that the geometric mean of any finite subset of S is an integer?

1984/3. (G.I./9). P, A, B, C and D are five distinct points in space such that $\angle APB = \angle BPC = \angle CPD = \angle DPA = \theta$, where θ is a given acute angle. Determine the greatest and least values of $\angle APC + \angle BPD$.

1984/4. (C. & P./10). A difficult mathematical competition consisted of a Part I and a Part II with a combined total of 28 problems. Each contestant solved 7 problems altogether. For each pair of problems, there were exactly two contestants who solved both of them. Prove that there was a contestant who, in Part I, solved either no problems or at least four problems.

1984/5. (A/13). $P(x)$ is a polynomial of degree $3n$ such that
$$P(0) = P(3) = \cdots = P(3n) = 2,$$
$$P(1) = P(4) = \cdots = P(3n - 2) = 1,$$
$$P(2) = P(5) = \cdots = P(3n - 1) = 0, \quad \text{and}$$
$$P(3n + 1) = 730.$$
Determine n.

Fourteenth U.S.A. Mathematical Olympiad, April 30, 1985

1985/1. (N.T./4). Determine whether or not there are any positive integral solutions of the simultaneous equations
$$x_1^2 + x_2^2 + \cdots + x_{1985}^2 = y^3,$$
$$x_1^3 + x_2^3 + \cdots + x_{1985}^3 = z^2$$
with distinct integers $x_1, x_2, \ldots, x_{1985}$.

1985/2. (I/7). Determine each real root of

$$x^4 - (2 \cdot 10^{10} + 1)x^2 - x + 10^{20} + 10^{10} - 1 = 0$$

correct to four decimal places.

1985/3. (G.I./8). Let A, B, C and D denote any four points in space such that at most one of the distances AB, AC, AD, BC, BD and CD is greater than 1. Determine the maximum value of the sum of the six distances.

1985/4. (C. & P./4). There are n people at a party. Prove that there are two people such that, of the remaining $n - 2$ people, there are at least $\lfloor n/2 \rfloor - 1$ of them, each of whom knows both or else knows neither of the two. Assume that "knowing" is a symmetrical relation; $\lfloor x \rfloor$ denotes the greatest integer less than or equal x.

1985/5. (C. & P./9). Let $a_1, a_2, a_3, \cdots$ be a non-decreasing sequence of positive integers. For $m \geqslant 1$, define $b_m = \min\{n: a_n \geqslant m\}$, that is, b_m is the minimum value of n such that $a_n \geqslant m$. If $a_{19} = 85$, determine the maximum value of

$$a_1 + a_2 + \cdots + a_{19} + b_1 + b_2 + \cdots + b_{85}.$$

Fifteenth U.S.A. Mathematical Olympiad, April 22, 1986

1986/1. (N.T./12).
(a) Do there exist 14 consecutive positive integers each of which is divisible by one or more primes p from the interval $2 \leqslant p \leqslant 11$?
(b) Do there exist 21 consecutive positive integers each of which is divisible by one or more primes p from the interval $2 \leqslant p \leqslant 13$?

1986/2. (C. & P./5). During a certain lecture, each of five mathematicians fell asleep exactly twice. For each pair of these mathematicians, there was some moment when both were sleeping simultaneously. Prove that, at some moment, some three were sleeping simultaneously.

1986/3. (N.T./8). What is the smallest integer n, greater than one, for which the root-mean-square of the first n positive integers is an integer?

Note. The root-mean square of n numbers $a_1, a_2, \ldots, a_n$ is defined to be

$$\left[\frac{a_1^2 + a_2^2 + \cdots + a_n^2}{n} \right]^{1/2}$$

1986/4. (P.G./5). Two distinct circles K_1 and K_2 are drawn in the plane. They intersect at points A and B, where AB is a diameter of K_1. A point P on K_2 and inside K_1 is also given.

Using only a "T-square" (i.e., an instrument which can produce the straight line joining two points and the perpendicular to a line through a point on or off the line), find a construction for two points C and D on K_1 such that CD is perpendicular to AB and CPD is a right angle.

1986/5. (N.T./13). By a *partition π of an integer $n \geq 1$*, we mean here a representation of n as a sum of one or more positive integers where the summands must be put in nondecreasing order. (E.g., if $n = 4$, then the partitions π are $1 + 1 + 1 + 1$, $1 + 1 + 2$, $1 + 3$, $2 + 2$, and 4).

For any partition π, define $A(\pi)$ to be the number of 1's which appear in π, and define $B(\pi)$ to be the number of distinct integers which appear in π. (E.g., if $n = 13$ and π is the partition $1 + 1 + 2 + 2 + 2 + 5$, then $A(\pi) = 2$ and $B(\pi) = 3$).

Prove that, for any fixed n, the sum of $A(\pi)$ over all partitions π of n is equal to the sum of $B(\pi)$ over all partitions π of n.

Solutions

Algebra

A/1. (1980/1). A two-pan balance is inaccurate since its balance arms are of different lengths and its pans are of different weights. Three objects of different weights A, B, and C are each weighed separately. When placed on the left-hand pan, they are balanced by weights A_1, B_1, C_1, respectively. When A and B are placed on the right-hand pan, they are balanced by A_2 and B_2, respectively. Determine the true weight of C in terms of A_1, B_1, C_1, A_2 and B_2.

Solution. Take the length of the left balance arm as our unit of length, denote the length of the right balance arm by k, and let L and R denote the weights of the left and right hand pans, respectively, as in the figure.

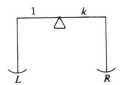

In order to balance, the sum of the (signed) moments of the weights about the fulcrum must $= 0$. This leads to the five equations:

$$(1) \quad A + L = k(A_1 + R), \quad (2) \quad B + L = k(B_1 + R),$$
$$(3) \quad C + L = k(C_1 + R), \quad (4) \quad A_2 + L = k(A + R),$$
$$(5) \quad B_2 + L = k(B + R).$$

Equations (1) and (2) imply that
$$(A - B) = k(A_1 - B_1);$$
equations (4) and (5) imply that
$$(A_2 - B_2) = k(A - B).$$

Thus,

$$k^2 = (A_2 - B_2)/(A_1 - B_1).$$

Subtracting (1) from (4), and solving for A, we get

$$A = \frac{k(A_1 + A_2)}{k + 1};$$

subtracting (1) from (3), and solving for C, we get

$$C = A + k(C_1 - A_1).$$

Finally, substituting for k and A, we get after some algebraic simplification that

$$C = \frac{C_1\{\sqrt{(A_1 - B_1)(A_2 - B_2)} + (A_2 - B_2)\} + A_1 B_2 - A_2 B_1}{\sqrt{(A_1 - B_1)(A_2 - B_2)} + A_1 - B_1}.$$

A/2. (1983/2). Prove that the roots of

$$x^5 + ax^4 + bx^3 + cx^2 + dx + e = 0$$

cannot *all* be real if $2a^2 < 5b$.

Solution. Denoting the roots by r_i, $i = 1, 2, \ldots, 5$, we have

$$-a = \sum r_i \quad \text{and} \quad b = \sum r_i r_j.$$

Then $2a^2 - 5b < 0$ is equivalent to

$$2\{\sum r_i\}^2 - 5\sum r_i r_j < 0 \quad \text{or} \quad \sum (r_i - r_j)^2 < 0.$$

Therefore, all the roots cannot be real.

One of the contestants used a calculus proof via Rolles's theorem. In order for $F(x) = x^5 + ax^4 + bx^3 + cx^2 + dx + e$ to have five real zeros, its derivative $F'(x)$ must have four real zeros. Then similarly, $F''(x)$ and $F'''(x)$ must have three and two real zeros, respectively. However, the discriminant of the quadratic $F'''(x) = 60x^2 + 24ax + 6b$ is $288(2a^2 - 5b)$, which is negative by hypothesis. Thus $F(x) = 0$ cannot have all real roots.

Exercise: Show that the roots of the equation

$$x^n + a_1 x^{n-1} + \cdots + a_{n-1} x + a_n = 0$$

cannot *all* be real if $(n - 1)a_1^2 - 2na_2 < 0$. (Letting $n = 5$, we recapture the above result).

A/3. (1984/1). The product of two of the four roots of the quartic equation $x^4 - 18x^3 + kx^2 + 200x - 1984 = 0$ is -32. Determine the value of k.

Solution. If r_1, r_2, r_3, and r_4 are the four roots, then for some pairing of the roots, $r_1r_2 = -32$, and then

$$r_3r_4 = \frac{r_1r_2r_3r_4}{r_1r_2} = \frac{-1984}{-32} = 62.$$

Consequently, for some p and q,

$$x^4 - 18x^3 + kx^2 + 200x - 1984 \equiv (x - r_1)(x - r_2)(x - r_3)(x - r_4)$$
$$= (x^2 - px - 32)(x^2 - qx + 62).$$

Equating like coefficients on both sides of the identity, we find

$$p + q = 18, \qquad -62p + 32q = 200, \qquad \text{and} \quad k = 62 + pq - 32.$$

Solving the first two equations for p, q, we get $p = 4$, $q = 14$. Finally, $k = 62 + 4 \cdot 14 - 32 = 86$.

A/4. (1977/3). If a and b are two of the roots of $x^4 + x^3 - 1 = 0$, prove that ab is a root of $x^6 + x^4 + x^3 - x^2 - 1 = 0$.

Solution. Let a, b, c, d be the roots of the given quartic equation, and also let $p = a + b$, $q = ab$, $r = c + d$, and $s = cd$. Then,

(1) $\qquad -1 = a + b + c + d = p + r,$

(2) $\qquad 0 = ab + ac + ad + bc + bd + cd = pr + q + s,$

(3) $\qquad 0 = abc + bcd + cda + dab = ps + qr,$

(4) $\qquad -1 = abcd = qs.$

We now eliminate p, r, and s from these equations. From (1) and (4), $r = -1 - p$, $s = -1/q$ and substituting these values in (2) and (3), gives

$$p(1 + p) = q - \frac{1}{q} \qquad \text{and} \qquad p = \frac{-q^2}{q^2 + 1}.$$

Finally,

$$\frac{-q^2}{(q^2 + 1)^2} = \frac{q^2 - 1}{q} \qquad \text{or} \qquad q^6 + q^4 + q^3 - q^2 - 1 = 0.$$

More generally if a, b, c, d are roots of $x^4 - a_1x^3 + a_2x^2 - a_3x + a_4 = 0$, we can determine the 6th degree polynomial equation of which ab

(or any of the 5 other products ac, bd, etc.) is a root. We just replace the left hand sides of (1) through (4) by a_1, a_2, a_3 and a_4, respectively, and proceed as before to give (after some manipulation) that

$$x^6 - a_2x^5 + (a_1a_3 - a_4)x^4 + (2a_2a_4 - a_3^2 - a_1^2a_4)x^3$$
$$+ (a_1a_3 - a_4)a_4x^2 - a_2a_4^2x + a_4^3 = 0.$$

Note. In principle, one could take any rational function of the roots, e.g., $a/(b^2 + c^3)$, and find a polynomial equation with integer coefficients it satisfies. The problem amounts to expressing the elementary symmetric functions of $a/(b^2 + c^3)$ etc., in terms of the elementary symmetric functions of a, b, c, d which are a_1, a_2, a_3, and a_4. For this case there will be a considerable amount of algebra since the resulting polynomial will be of degree $3! \cdot \binom{4}{3} = 24$ corresponding to the various possibilities $a/(b^2 + c^3), a/(b^3 + c^2), b/(c^2 + a^3), b/(c^3 + a^2)$, etc. of the roots.

A/5. (1973/4). Determine all the roots, real or complex, of the system of simultaneous equations

$$x + y + z = 3,$$
$$x^2 + y^2 + z^2 = 3,$$
$$x^3 + y^3 + z^3 = 3.$$

Solution. Let x, y, z be the roots of the cubic equation

$$t^3 - at^2 + bt - c = 0.$$

Then,

$$a = x + y + z = 3,$$

and

$$2b = 2(yz + zx + xy) = (x + y + z)^2 - x^2 - y^2 - z^2 = 6.$$

The cubic is satisfied by x, by y and by z; if we sum these cubics we obtain

$$x^3 + y^3 + z^3 - a(x^2 + y^2 + z^2) + b(x + y + z) - 3c = 0$$
$$3 - 3a + 3b - 3c = 0.$$

Since $a = b = 3$, we find $c = 1$; the cubic equation becomes $(t - 1)^3 = 0$, and so the only solution is $(x, y, z) = (1, 1, 1)$.

More generally we show that if

$$\sum_{i=1}^{n} x_i^k = \sum_{i=1}^{n} a_i^k, \qquad k = 1, 2, \ldots, n,$$

and the a_i are given constants, then

$$(x_1, x_2, \ldots, x_n) = (a_1, a_2, \ldots, a_n)$$

aside from permutations.†

Our proof uses the Newton formulas. With the abbreviations

$$S_r = \sum a_i^r \quad \text{and} \quad T_1 = \sum x_i, \quad T_2 = \sum x_i x_j, \quad T_3 = \sum x_i x_j x_k, \ldots,$$

they state

$$S_1 - T_1 = 0, \qquad S_2 - T_1 S_1 + 2T_2 = 0, \qquad \ldots,$$

$$S_{n-1} - T_1 S_{n-2} + T_2 S_{n-3} + \cdots + (-1)^{n-1}(n-1)T_{n-1} = 0,$$

and

$$S_r - T_1 S_{r-1} + T_2 S_{r-2} + \cdots + (-1)^n T_n S_{r-n} = 0, \quad \text{for } r \geq n.$$

From these it follows that S_k is determined uniquely as a function of $T_1, T_2, \ldots, T_k$ and also that T_k is determined uniquely as a function of $S_1, S_2, \ldots, S_k$. Consequently, the elementary symmetric functions (the T_i's) of the x_i's must be identical to the elementary symmetric functions of the a_i's. Whence

$$\prod_{i=1}^{n} (x - x_i) = \prod_{i=1}^{n} (x - a_i),$$

which gives the desired result.

The original problem corresponds to the special case $n = 3$, and $a_1 = a_2 = a_3 = 1$.

A/6. (1977/1). Determine all pairs of positive integers (m, n) such that

$$(1 + x^n + x^{2n} + \cdots + x^{mn}) \quad \text{is divisible by} \quad (1 + x + x^2 + \cdots + x^m).$$

Solution. Since the given expressions are equivalent to

$$\frac{x^{n(m+1)} - 1}{x^n - 1} \quad \text{and} \quad \frac{x^{m+1} - 1}{x - 1},$$

respectively,

$$\frac{(x^{n(m+1)} - 1)(x - 1)}{(x^{m+1} - 1)(x^n - 1)}$$

must be a polynomial. Since all the factors of $x^{n(m+1)} - 1$ are distinct,

---

†See M. S. Klamkin and D. J. Newman, Uniqueness theorems for power equations, Elem. der Mathematik 25/6 (1970) 130–134.

$x^{m+1} - 1$ and $x^n - 1$ cannot have any common factors other than $x - 1$. Thus $m + 1$ and n must be relatively prime. This condition is also sufficient since

$$x^{n(m+1)} - 1 = (x^n)^{m+1} - 1 = (x^{m+1})^n - 1$$

and is therefore divisible by $x^n - 1$ and by $x^{m+1} - 1$.

A/7. (1974/1). Let a, b, and c denote three distinct integers, and let P denote a polynomial having all integral coefficients. Show that it is impossible that $P(a) = b$, $P(b) = c$, and $P(c) = a$.

Solution. More generally we will show that if $a_1, a_2, \ldots, a_n$ are distinct integers and P is a polynomial with integer coefficients, then it is impossible to have $P(a_k) = a_{k+1}$ for $k = 1, 2, \ldots, n$ (here $a_{n+k} = a_k$).

Since P is an integral polynomial $b_0 + b_1 x + \cdots + b_m x^m$, it follows that for any two integers r, s

$$P(r) - P(s) = (r - s)Q(r, s),$$

where $Q(r, s)$ is an integer. Consequently, if $P(a_k) = a_{k+1}$ for all k, we would have

$$a_{k+1} - a_{k+2} = P(a_k) - P(a_{k+1}) = (a_k - a_{k+1})Q(a_k, a_{k+1})$$

for $k = 1, 2, \ldots, n$. Taking the product of these n equations we find that

$$Q(a_1, a_2)Q(a_2, a_3) \cdots Q(a_n, a_1) = 1.$$

Thus $Q(a_k, a_{k+1}) = \pm 1$ for all k, and so $a_{k+1} - a_{k+2} = \pm(a_k - a_{k+1})$, and

$$|a_1 - a_2| = |a_2 - a_3| = \cdots = |a_n - a_1|.$$

From $|a_1 - a_2| = |a_2 - a_3|$ we conclude that

$$a_1 - a_2 = a_2 - a_3 \quad \text{or} \quad a_1 - a_2 = -(a_2 - a_3).$$

The second of these gives $a_1 = a_3$, which contradicts the assumption that the a_i are distinct. Hence $a_1 - a_2 = a_2 - a_3$ and similarly

$$a_1 - a_2 = a_2 - a_3 = \cdots = a_n - a_1.$$

This is impossible since the sum of these n latter numbers is zero, and this can only be if all the a_i's were the same.

A/8. (1975/3). If $P(x)$ denotes a polynomial of degree n such that $P(k) = k/(k + 1)$ for $k = 0, 1, 2, \ldots, n$, determine $P(n + 1)$.

Solution. Let $Q(x) = (x + 1)P(x) - x$. Then, since $Q(x)$ is a polynomial of $(n + 1)$-st degree and vanishes for $x = 0, 1, 2, \ldots, n$, we must have

$$Q(x) = (x + 1)P(x) - x = Ax(x - 1)(x - 2) \cdots (x - n),$$

where A is a constant which can be determined by setting $x = -1$, i.e., $1 = A(-1)^{n+1}(n + 1)!$ Then

$$P(x) = \frac{1}{x + 1}\left(x + \frac{(-1)^{n+1}x(x - 1) \cdots (x - n)}{(n + 1)!}\right)$$

Finally,

$$P(n + 1) = 1 \quad \text{or} \quad \frac{n}{n + 2}, \quad \text{according as } n \text{ is odd or even.}$$

A/9. (1976/5). If $P(x)$, $Q(x)$, $R(x)$, and $S(x)$ are all polynomials such that

$$P(x^5) + xQ(x^5) + x^2R(x^5) = (x^4 + x^3 + x^2 + x + 1)S(x),$$

prove that $x - 1$ is a factor of $P(x)$.

Solution. More generally, we show that if $P_0(x), P_1(x), \ldots, P_{n-2}(x)$ $(n \geqslant 2)$ and $S(x)$ are polynomials such that

$$P_0(x^n) + xP_1(x^n) + \cdots + x^{n-2}P_{n-2}(x^n)$$
$$= (x^{n-1} + x^{n-2} + \cdots + x + 1)S(x),$$

then $x - 1$ is a factor of $P_i(x)$ for all i.

Let ω_i, $i = 1, 2, \ldots, n - 1$ denote the complex n^{th} roots of unity other than 1. Since

$$(x^n - 1) = (x - 1)(1 + x + \cdots + x^{n-1}),$$

it follows that

$$1 + \omega_i + \omega_i^2 + \cdots + \omega_i^{n-1} = 0$$

for all i. If we now substitute ω_i in the given identity above, we find that

$$P_0(1) + \omega_i P_1(1) + \cdots + \omega_i^{n-2}P_{n-2}(1) = 0$$

for all i. Since the $(n - 2)^{\text{th}}$ degree polynomial

$$P_0(1) + xP_1(1) + \cdots + x^{n-2}P_{n-2}(1)$$

has $n - 1$ distinct roots ω_i, it must vanish identically. Thus,

$$P_0(1) = P_1(1) = \cdots = P_{n-2}(1) = 0.$$

Finally, by the Factor Theorem, $x - 1$ is a factor of each of $P_0(x)$, $P_1(x), \ldots, P_{n-2}(x)$.

A/10. (1973/2). Let $\{X_n\}$ and $\{Y_n\}$ denote two sequences of integers defined as follows:

$$X_0 = 1, \quad X_1 = 1, \quad X_{n+1} = X_n + 2X_{n-1} \quad (n = 1, 2, 3, \ldots),$$

$$Y_0 = 1, \quad Y_1 = 7, \quad Y_{n+1} = 2Y_n + 3Y_{n-1} \quad (n = 1, 2, 3, \ldots).$$

Thus, the first few terms of the sequences are:

$$X: \quad 1, 1, \ 3, \ 5, \ 11, \ 21, \ldots,$$

$$Y: \quad 1, 7, 17, 55, 161, 487, \ldots .$$

Prove that, except for the "1", there is no term which occurs in both sequences.

Solution.   Mod 8, the first few terms of the two sequences are

$$X: \quad 1, 1, 3, 5, 3, 5, \ldots,$$

$$Y: \quad 1, 7, 1, 7, 1, 7, \ldots .$$

An easy induction shows that this alternate periodic behaviour persists. Thus 1 is the only common term of the two sequences.

A more general approach involves solving the given difference equations. The theory states† that if a, b, x_0, x_1 are given numbers and $x_2, x_3, \ldots$ are determined recursively by means of

$$x_{n+1} = ax_n + bx_{n-1}, \qquad\qquad n = 1, 2, 3, \ldots,$$

and if $a^2 + 4b \neq 0$, then x_n can be expressed in terms of a, b, x_0, x_1 by the formula

$$x_n = \frac{(x_1 - k_1 x_0)k_2^n - (x_1 - k_2 x_0)k_1^n}{k_2 - k_1},$$

where k_1, k_2 are the roots of $k^2 - ak - b = 0$. (If $a^2 + 4b = 0$, $k_1 = k_2$ and the formula for x_n is different.)

In the case of the first sequence in this problem, we have $x_0 = 1$, $x_1 = 1$, $a = 1$, $b = 2$ and find $k_2 = 2, k_1 = -1$ (or vice versa); we get

$$x_n = \tfrac{1}{3}\left[2^{n+1} + (-1)^n\right].$$

†See [60, pp. 367–370] or [126, pp. 121–122].

For the second sequence, we find

$$y_n = 2 \cdot 3^n - (-1)^n.$$

To get $x_n = y_m$, we must have

$$3^{m+1} - 2^n = \tfrac{1}{2}\left[3(-1)^m + (-1)^n\right].$$

If $n = 0$ or $n = 1$, we see that $m = 0$ is the only solution. Henceforth, take $n \geqslant 2$. If m and n are both even, or both odd, the right member of this equation is even, but the left member is odd. If m and n are of opposite parity, the equation is invalid $\bmod 4$.

A/11. (1980/3). Let $F_r = x^r \sin rA + y^r \sin rB + z^r \sin rC$, where x, y, z, A, B, C are real and $A + B + C$ is an integral multiple of π. Prove that if $F_1 = F_2 = 0$, then $F_r = 0$ for all positive integral r.

Solution. Let

$$u = x(\cos A + i \sin A) = xe^{iA}, \qquad v = y(\cos B + i \sin B) = ye^{iB},$$
$$w = z(\cos C + i \sin C) = ze^{iC}$$

and

$$G_r = u^r + v^r + w^r.$$

Thus $G_0 = 3$ and $G_r = H_r + iF_r$, where

$$H_r = x^r \cos rA + y^r \cos rB + z^r \cos rC.$$

If $F_1 = F_2 = 0$, then G_1 and G_2 are real, and it now suffices to show that G_r is real for $r = 3, 4, \ldots$.

First consider the cubic equation $p(x) = x^3 - ax^2 + bx - c = 0$ whose roots are u, v, w. We have

$$a = u + v + w = G_1,$$

$$2b = 2(vw + wu + uv) = (u + v + w)^2 - (u^2 + v^2 + w^2) = G_1^2 - G_2.$$

By hypothesis, $A + B + C = k\pi$, k an integer; so

$$c = uvw = xyze^{ik\pi} = \pm xyz = \text{real}.$$

We now show inductively that G_r is always real, or equivalently that F_r is always 0, by obtaining a recurrence relation on the G_r. We multiply $p(x)$ by x^n and obtain

$$x^{3+n} - ax^{2+n} + bx^{1+n} - cx^n = 0, \quad \text{for } x = u, v, w.$$

Summing over the roots u, v, w, we get

$$G_{3+n} - aG_{2+n} + bG_{1+n} - cG_n = 0.$$

Since $G_0(= 3)$, G_1, G_2 and c are real, so is G_3; and then by induction, so are $G_4, G_5, \ldots$.

Alternate solution.   While this solution is not as elegant as the preceding one, it has the virtue of being very direct. We distinguish three cases.

Case (i) $\sin A \cdot \sin B \cdot \sin C \neq 0$.

We solve the equation $F_1 = 0$ for z:

(1)
$$z = \frac{x \sin A + y \sin B}{\sin C};$$

and substitute this into $F_2 = 0$:

$$x^2 \sin 2A + y^2 \sin 2B + \frac{(x \sin A + y \sin B)^2}{\sin^2 C} \sin 2C = 0.$$

Using the given information that $A + B + C = k\pi$, k an integer, we get after some algebraic and trigonometric simplification that

$$\sin A \sin B \left[x^2 + y^2 - (-1)^k 2xy \cos C \right] = 0.$$

Since $\sin A$ and $\sin B$ are $\neq 0$, we conclude that

(2)
$$x^2 + y^2 - 2(-1)^k xy \cos C = 0.$$

Since $\sin C \neq 0$, $|\cos C| < 1$. This implies that the quadratic form on the left in (2) is positive definite; that is, it is ≥ 0 for all x, y and $= 0$ only for

$$x = y = 0.$$

[To see that the form is positive definite, either examine its discriminant or write it as the sum of squares $\left(x - (-1)^k y \cos C \right)^2 + (y \sin C)^2$.]

Now by (1), also $z = 0$. This shows that $F_r = 0$ for *all real r*, not only integers.

Case (ii) Exactly one of $\sin A, \sin B, \cos C$ vanishes.

Say $\sin A = 0$. Then A is an integer multiple of π and $\sin rA = 0$ for all integers r. Since $B + C = n\pi$, we have $\sin B + (-1)^n \sin C = 0$, and

(3)
$$\sin rB + (-1)^{rn} \sin rC = 0.$$

The condition $F_1 = 0$ is

$$F_1 = y \sin B + z \sin C = \left(y - (-1)^n z \right) \sin B = 0.$$

Since $\sin B \neq 0$, we conclude that

(4)
$$y = (-1)^n z.$$

Then from (3) and (4)

$$F_r = y^r \sin rB + z^r \sin rC$$
$$= y^r \sin rB - (-1)^{rn} y^r (-1)^{rn} \sin rB = 0$$

for all integers r.

Note that we used only $F_1 = 0$.

Case (iii) $\sin A = \sin B = \sin C = 0$

Clearly $F_r = 0$ for all integers, with no assumptions.

For a generalization, let $F_r = \Sigma x_i^r \sin rA_i$ where x_i, A_i, $i = 1, 2, \ldots, n$ are real, $\Sigma A_i = m\pi$ (m integer). If $F_1 = F_2 = \cdots = F_{n-1} = 0$, then $F_r = 0$ for all positive integral r. The proof is similar to the one we gave first. Letting $u_j = x_j e^{iA_j}$, $G_r = u_j^r$, so that $G_1, G_2, \ldots, G_{n-1}$ are real, all we need show is that G_r is real for all r. We now consider the nth degree equation

$$(5) \qquad x^n - a_1 x^{n-1} + a_2 x^{n-2} - \cdots + (-1)^n a_n = 0,$$

whose roots are $u_1, u_2, \ldots, u_n$. Then since $u_1 u_2 \ldots u_n$ is also real, it follows from the Newton formulae (see A/5), that the a_j's are all real. Then summing over the roots of (5), we find inductively that G_r is real for all r.

A/12. (1982/2). Let $S_r = x^r + y^r + z^r$ with x, y, z real. It is known that if $S_1 = 0$,

$$(*) \qquad \frac{S_{m+n}}{m+n} = \frac{S_m}{m} \frac{S_n}{n}$$

for $(m, n) = (2, 3), (3, 2), (2, 5),$ or $(5, 2)$. Determine *all* other pairs of integers (m, n) if any, so that $(*)$ holds for all real numbers x, y, z such that $x + y + z = 0$.

Solution. We will show that $(*)$ does not hold for any other pairs of integers.

(a) Firstly, $m, n > 0$ otherwise S_m or S_n will not be defined for, say, $x = 0$.

(b) m and n cannot both be odd; otherwise, by letting $(x, y, z) = (1, -1, 0)$, we would have $S_{m+n} = 2$, $S_m = 0 = S_n$ which violates $(*)$.

(c) m and n cannot both be even. This follows again by letting $(x, y, z) = (1, -1, 0)$ which gives $S_{m+n} = S_m = S_n = 2$. Then $(*)$ becomes $2/(m+n) = (2/m)(2/n)$ or $(m-2)(n-2) = 4$, so that $m = n = 4$. But then $S_8/8 \neq (S_4/4)(S_4/4)$.

(d) We now assume m and n are of opposite parity, say m odd and n even. In the case $n = 2$, $(*)$ with $(x, y, z) = (-1, -1, 2)$ reduces to

$$(m - 6)2^m = -4m - 12$$

after some simple algebra. The latter equation holds for $m = 3$ and 5, in conformity with the statement of the problem. For $m \geqslant 7$ there are no solutions since then the two sides are of opposite sign. In the case $n \geqslant 4$, we show that $(*)$ is impossible with $(x, y, z) = (-1, -1, 2)$. Here $(*)$ simplifies to

$$(1) \qquad (2^{m+n} - 2)(mn - m - n) = (m + n)(2^{m+1} - 2^{n+1} - 2).$$

Since $mn - m - n > 0$, each of the four expressions in the parentheses must be positive. For the expression in the last of the four parentheses to be positive we must have $m > n$, so that $m \geqslant 5$. Since

$$(m + n)(2^{m+1} - 2^{n+1} - 2) < (m + n)(2^{m+n} - 2),$$

it follows from (1) that $mn - m - n < m + n$ or that $(m - 2)(n - 2) < 4$. This is impossible, since $m \geqslant 5$ and $n \geqslant 4$.

The problem is more involved if we add the requirement that $xyz \neq 0$, since then negative values of m, n must be considered. The previous pairs for (m, n) are still valid, and an analysis similar to that above, but longer, shows that the only other possible solutions of $(*)$ are $(m, n) = (3, -1)$ or $(-1, 3)$, that is

$$(2) \qquad \frac{x^2 + y^2 + z^2}{2} = \frac{x^3 + y^3 + z^3}{3} \cdot \frac{x^{-1} + y^{-1} + z^{-1}}{-1}$$

holds whenever $x + y + z = 0$ and $xyz \neq 0$. To verify that (2) is indeed true, just note that

$$\sum x^2 = -2 \sum yz, \qquad \sum x^3 = 3xyz, \qquad \text{and} \qquad \sum x^{-1} = \sum (yz/xyz),$$

where the sums are cyclic over x, y, z.

We now give an alternative interesting algebraic solution by Peter Lax.

Eliminate z by setting $z = -(x + y)$, and set $x/y = \xi$. Then

$$(1)' \qquad S_r(x, y, z) = y^r P_r(\xi),$$

where $P_r(\xi)$ is a polynomial given by

$$(2)' \qquad P_r(\xi) = \xi^r + 1 + (-1)^r (1 + \xi)^r.$$

S_r is invariant under permutations of x, y, z. It follows that P_r is invariant under

$$(3) \qquad \begin{aligned} P(\xi) &\to P(-\xi - 1), \\ P(\xi) &\to \xi^r P(1/\xi), \end{aligned}$$

and of course under any composition of these maps.

Therefore, if P is invariant under the maps (3), and if ρ is a root of P, then so are $-1 - \rho$ and $1/\rho$. That is, the set of roots is invariant under the mappings

$$(4) \qquad \begin{aligned} &\rho \to -\rho - 1, \quad \rho \to \frac{1}{\rho}, \quad \rho \to -\frac{1}{\rho} - 1, \\ &\rho \to -\frac{1}{\rho + 1}, \quad \rho \to -\frac{\rho}{1 + \rho}. \end{aligned}$$

The last three are obtained from the first two by composition; these mappings, together with the identity, form a group isomorphic with the group of permutations of 3 elements.

If P has degree < 6, it has less than 6 roots, so any root ρ must coincide with one of its images under (4), i.e. must satisfy at least one of

$$A: \quad \rho = -\rho - 1, \qquad \rho = -1/2$$
$$B: \quad \rho = 1/\rho, \qquad \rho = 1 \text{ or } \rho = -1$$
$$C: \quad \rho = \frac{1}{\rho} - 1, \qquad \rho^2 + \rho + 1 = 0$$
$$D: \quad \rho = -\frac{1}{\rho + 1}, \qquad \rho^2 + \rho + 1 = 0$$
$$E: \quad \rho = -\frac{\rho}{1 + \rho}, \qquad \rho = 0.$$

In case A, $\{-1/2, -2, 1\}$ are among the zeros, as they are in case B_+. The polynomial with these zeros,

$$Q_1(\xi) = (2\xi + 1)(\xi + 2)(\xi - 1) = 2\xi^3 + 3\xi^2 - 3\xi - 2$$

is invariant under the mappings (3). So is

$$Q_2(\xi) = \xi^2 + \xi + 1,$$

corresponding to cases C and D, and

$$Q_3(\xi) = \xi^2 + \xi,$$

corresponding to cases B_- and E.

It follows that any invariant polynomial P of degree < 6 has a zero in common with Q_1, Q_2 or Q_3, and therefore has all its zeros. Thus such a P is divisible by Q_1, Q_2 or Q_3. Since the quotient P/Q_i is also invariant and of degree < 6, P is, in fact, a product of the Q_i. We note also that an invariant P is divisible by Q_3 if and only if it vanishes at $\xi = 0$.

We compute P_r from (2) for $r = 2, 3, 4, 5$, and find, indeed, that

$$(5) \qquad P_2 = 2Q_2, \quad P_3 = -3Q_3, \quad P_4 = 2Q_2^2, \quad P_5 = -5Q_3Q_2.$$

P_6 is of degree 6 and is not divisible by any Q. The polynomial

$$P_7(\xi) = -7(\xi^6 + 3\xi^5 + 5\xi^4 + 5\xi^3 + 3\xi^2 + \xi)$$

vanishes at $\xi = 0$ and at $\xi = -1$, and thus contains all zeros of Q_3 and is divisible by it. The quotient P_7/Q_3 is of degree 4, does not vanish at $\xi = 0$, so it must be a constant multiple of the square of Q_2:

$$(5)' \qquad P_7 = -7Q_3Q_2^2.$$

Relations (5) and (5)' imply all the listed identities for the S_r.

Suppose P is invariant, of degree n; denote its roots by $\rho_1, \ldots, \rho_n$. The transformations (4) permute these, and thus divide them into groups of 6's, 3's or 2's. In particular, if $n \equiv 2 \pmod 6$, there must be at least one group of 2's, and if $n \equiv 4 \pmod 6$, at least 2 groups of 2's and therefore P is divisible by Q_2 or Q_3.

We apply these observations to P_r. Since for r even, P_r is of degree r and does not vanish at $\xi = 0$, we conclude, using (5):

$$(6) \qquad \begin{array}{l} \text{For } r \equiv 2 \pmod 6, \ P_r \text{ is divisible by } \tfrac{1}{2}P_2. \\ \text{For } r \equiv 4 \pmod 6, \ P_r \text{ is divisible by } \tfrac{1}{4}P_2^2. \end{array}$$

For r odd, P_r is of degree $r - 1$ and has $\rho = 0$ as simple zero; thus it is divisible by Q_3. We conclude, using (5) and (5)':

$$\text{For } r \equiv 3 \pmod 6, \ P_r \text{ is divisible by } \tfrac{1}{3}P_3.$$

(6)'   For $r \equiv 5 \pmod 6$ P_r is divisible by $\tfrac{1}{5}P_5$.

$$\text{For } r \equiv 1 \pmod 6 \ P_r \text{ is divisible by } \tfrac{1}{7}P_7.$$

Note that P_j/j has integer coefficients and leading coefficient ± 1 for $j = 2, 3, 5, 7$. Since P_r has integer coefficients, it follows that if P_r is divisible by P_j/j, the quotient again has integer coefficients.

We combine the above observation with (6) and (6)'; using (1)' relating P_r to S_r, we deduce the following divisibility properties of integers:

Take x, y, z to be integers, $x + y + z = 0$. Then $S_r(x, y, z)$ is divisible by $\tfrac{1}{2}S_2(x, y, z)$ when $r \equiv 2 \pmod 6$, and by $\tfrac{1}{4}S_2^2(x, y, z)$ when $r \equiv 4 \pmod 6$. When $r \equiv 3, 5$ or $1 \pmod 6$, $S_r(x, y, z)$ is divisible by $\tfrac{1}{3}S_3(x, y, z)$, $\tfrac{1}{5}S_5(x, y, z)$ or $\tfrac{1}{7}S_7(x, y, z)$, respectively.

Exercise 1: Show that if $P(\xi)$ is a polynomial of degree 6, invariant under

(7) $P(\xi) \to P(-\xi - 1), \qquad P(\xi) \to \xi^6 P(1/\xi),$

and if the leading coefficient of P is 1, then P is of the form

(7)' $P(\xi) = \xi^6 + 3\xi^5 + a\xi^4 + (2a - 5)\xi^3 + a\xi^2 + 3\xi + 1.$

Conversely, show that every P of the above form (7)' is invariant in the sense of (7).

Exercise 2: Compute P_{12} from (2) and factor it as 2 times two polynomials of form (7)'.

Comment. Note that the divisibility results of (6) and (6)' above follow directly from the following formulae due to Loren C. Larson, Crux Mathematicorum 9 (1983) p. 284:

$$S_{2m} = \sum_{k=0}^{[m/3]} \frac{2m}{m - k} \binom{m - k}{2k} X^{m - 3k} Y^{2k}$$

$$S_{2m+1} = \sum_{k=0}^{[(m-1)/3]} \frac{2m + 1}{2k + 1} \binom{m - k - 1}{2k} X^{m - 3k - 1} Y^{2k + 1},$$

where $X = S_2/2$ and $Y = S_3/3$.

A/13. (1984/5). $P(x)$ is a polynomial of degree $3n$ such that

$$P(0) = P(3) = \cdots = P(3n) = 2,$$
$$P(1) = P(4) = \cdots = P(3n - 2) = 1,$$
$$P(2) = P(5) = \cdots = P(3n - 1) = 0, \quad \text{and}$$
$$P(3n + 1) = 730.$$

Determine n.

Solution. Note that $P(x) - 1$ runs cyclically $1, 0, -1, 1, 0, -1, \ldots, 1$ as x goes from 0 to $3n$ in steps of 1. To mimic this behavior, consider the cube root of unity $\omega = (-1 + i\sqrt{3})/2$. Observe that

$$\{\omega^n\} = \{1, \omega, \omega^2, 1, \omega, \omega^2, \ldots\},$$

$$\{2\,\mathrm{Im}(\omega^n)/\sqrt{3}\} = \{0, 1, -1, 0, 1, -1, \ldots\}$$

and

$$\{2\,\mathrm{Im}(\omega^{2n+1})/\sqrt{3}\} = \{1, 0, -1, 1, 0, -1, \ldots\}.$$

Hence for $x = 0, 1, 2, \ldots, 3n$,

$$P(x) - 1 = 2\,\mathrm{Im}\,\omega^{2x+1}/\sqrt{3},$$

and by the binomial theorem

$*$ $$\omega^{2x} = \{1 + (\omega^2 - 1)\}^x = \sum_{k=0}^{x} \binom{x}{k}(\omega^2 - 1)^k.$$

Now define

$$Q(x) = \frac{2}{\sqrt{3}}\mathrm{Im}\left\{\omega \sum_{k=0}^{3n} \binom{x}{k}(\omega^2 - 1)^k\right\}.$$

Note that $\binom{x}{k} = 0$ if x is an integer less than k, so that $Q(x)$ agrees with $P(x) - 1$ for $x = 0, 1, 2, \ldots, 3n$, a total of $3n + 1$ values. Since $P(x) - 1$ and $Q(x)$ are both of degree $3n$, we have $P(x) - 1 \equiv Q(x)$. Using $(*)$ for $x = 3n + 1$, we obtain

$$P(3n + 1) - 1 = \frac{2}{\sqrt{3}}\mathrm{Im}\left\{\omega \sum_{k=0}^{3n} \binom{3n+1}{k}(\omega^2 - 1)^k\right\}$$

$$= \frac{2}{\sqrt{3}}\mathrm{Im}\left\{\omega\left(\omega^{2(3n+1)} - (\omega^2 - 1)^{3n+1}\right)\right\}.$$

Since $\mathrm{Im}(\omega^{6n+3}) = 0$, and $\omega^2 - 1 = i\omega\sqrt{3}$,

$$P(3n + 1) - 1 = -2\,\mathrm{Re}\left\{\omega^2(i\sqrt{3})^{3n}\right\}$$

$$= \begin{cases} (-1)^k 3^{3k} & \text{if } n = 2k, \\ (-1)^k 3^{3k+2} & \text{if } n = 2k + 1. \end{cases}$$

Finally, since $P(3n + 1) - 1 = 729 = 3^6$, we find that $n = 4$.

Remark: More generally, if $P(x)$ is cyclically equal to $a, b, c, a, b, c, \ldots a$ for $x = 0, 1, 2, \ldots, 3n$, then $P(x) = A + B\omega^x + C\omega^{2x}$ for $x = 0, 1, 2, \ldots, 3n$. The constants A, B and C are determinable by setting $x = 0, 1, 2$, and then we proceed as before using the two expansions

$$\omega^x = \{1 + (\omega - 1)\}^x \quad \text{and} \quad \omega^{2x} = \{1 + (\omega^2 - 1)\}^x.$$

Similarly, if $P(x)$ is cyclic of order r for $x = 0, 1, 2, \ldots, rn$, we would use the r-th roots of unity.

Alternate solution. Let $Q(x) = P(x) - 1$. The sum $\displaystyle\sum_{k=0}^{3n+1} (-1)^k \binom{3n+1}{k} Q(k)$ vanishes since it is the $(3n + 1)$th finite difference of a $(3n)$th degree polynomial. Solving for $Q(3n + 1)$, we get

$$Q(3n + 1) = \sum_{k=0}^{3n} (-1)^k \binom{3n+1}{k} Q(k) = \sum_{k=0}^{3n+1} A_k \binom{3n+1}{k},$$

where $A_k = 1, 0, -1, -1, 0, 1$ according to whether $k = 0, 1, 2, 3, 4,$ or 5 modulo 6. We can determine the last sum by the usual method of using the roots of unity. For example, if we knew that $F(x) = a_0 + a_1 x + \cdots + a_n x^n$ and we wanted the sum $a_0 + a_3 + a_6 + \cdots$, it would be given by $\frac{1}{3}[F(1) + F(\omega) + F(\omega^2)]$, where ω is a primitive cubic root of one. So starting from

$$(1 + x)^{3n+1} = \sum_{k=0}^{3n+1} \binom{3n+1}{k} x^k,$$

we find that

$$6Q(3n + 1) = \sum (1 + r_i)^{3n+1} \{ 1 - r_i^{-2} - r_i^{-3} + r_i^{-5} \}$$

where the sum is over the 6th roots of unity. Since $r_i^6 = 1$, we have

(1) $$6Q(3n + 1) = \sum (1 + r_i)^{3n+1} \{ 1 + r_i \} \{ 1 - r_i^3 \}.$$

The 6th roots of unity consist of three cube roots of 1 and three cube roots of -1. The sum over the cube roots of 1 in (1) is zero so that we get

$$3Q(3n + 1) = \left[(1 + r_1)^{3n+2} + (1 + r_2)^{3n+2} \right]$$

where $r_1 = e^{i\pi/3}$ and $r_2 = e^{-i\pi/3}$. Then since $1 + r_1 = 2e^{i\pi/6}\cos \pi/6$ and $1 + r_2 = 2e^{-i\pi/6}\cos \pi/6$,

$$3Q(3n + 1) = 3^{(3n+2)/2} \{ e^{i\pi(3n+2)/6} + e^{-i\pi(3n+2)/6} \}.$$

Finally,

$$3^7 = 3Q(3n + 1) = 2 \cdot 3^{(3n+2)/2}\cos(\pi n/2 + \pi/3),$$

so that n must be even and equal to 4.

Number Theory

N.T./1. (1972/1). The symbols $(a, b, \ldots, g)$ and $[a, b, \ldots, g]$ denote the greatest common divisor and least common multiple, respectively, of the positive integers $a, b, \ldots, g$. For example, $(3, 6, 18) = 3$ and $[6, 15] = 30$. Prove that

$$\frac{[a, b, c]^2}{[a, b][b, c][c, a]} = \frac{(a, b, c)^2}{(a, b)(b, c)(c, a)}.$$

Solution. Let

$$a = \prod p_i^{\alpha_i}, \quad b = \prod p_i^{\beta_i}, \quad c = \prod p_i^{\gamma_i},$$

where the p_i denote the prime factors of a, b, c (some of the exponents may be zero). Since

$$[a, b] = \prod p_i^{\max\{\alpha_i, \beta_i\}}, \quad (a, b) = \prod p_i^{\min\{\alpha_i, \beta_i\}},$$

etc., we have to show that

$$2\max\{\alpha_i, \beta_i, \gamma_i\} - \max\{\alpha_i, \beta_i\} - \max\{\beta_i, \gamma_i\} - \max\{\gamma_i, \alpha_i\}$$
$$= 2\min\{\alpha_i, \beta_i, \gamma_i\} - \min\{\alpha_i, \beta_i\} - \min\{\beta_i, \gamma_i\} - \min\{\gamma_i, \alpha_i\}$$

for each index i. Without loss of generality we can assume that $\alpha_i \geqslant \beta_i \geqslant \gamma_i$ for any particular index i. The equation to be established then reduces to the identity

$$2\alpha_i - \alpha_i - \beta_i - \alpha_i = 2\gamma_i - \beta_i - \gamma_i - \gamma_i.$$

The following interesting properties of the g.c.d. and the l.c.m. are noted in O. Ore, *Number Theory And Its History*, McGraw-Hill, N.Y., 1948, pp. 100–109:

1. *Idempotent law*:

$$(a, a) = a, \quad [a, a] = a.$$

2. *Commutative law*:

$$(a, b) = (b, a), \quad [a, b] = [b, a].$$

3. *Associative law*:

$$((a, b), c) = (a, (b, c)), \quad [[a, b], c] = [a, [b, c]].$$

4. *Absorption law*:

$$(a, [a, b]) = a, \quad [a, (a, b)] = a.$$

These laws follow from properties of the min and max operations, i.e.,

1. $\min(a, a) = a,$	$\max(a, a) = a.$
2. $\min(a, b) = \min(b, a),$	$\max(a, b) = \max(b, a)$
3. $\min\{\min(a, b), c\}$ $= \min\{a, \min(b, c)\},$	$\max\{\max(a, b), c\}$ $= \max\{a, \max(b, c)\}.$
4. $\min\{a, \max(a, b)\} = a,$	$\max\{a, \min(a, b)\} = a.$

From an examination of the above equations, it follows that g.c.d. and l.c.m., and min and max are dual operations. The conditions remain the same if g.c.d. is interchanged with l.c.m. or min with max. Also the set theory operations $\cup$ (union) and $\cap$ (intersection) satisfy the above laws. These operations are special cases of systems called *lattices* and which are treated in books on modern algebra.

Exercises: Using the method at the beginning, prove the following identities:

I_1. $(a, [b, c]) = [(a, b), (a, c)],$

$[a, (b, c)] = ([a, b], [a, c]).$

I_2. $([a, b], [b, c], [c, a]) = [(a, b), (b, c), (c, a)].$

I_3. $(ab, cd) = (a, c)(b, d)\left(\dfrac{a}{(a, c)}, \dfrac{d}{(b, d)}\right)\left(\dfrac{c}{(a, c)}, \dfrac{b}{(b, d)}\right).$

I_4. $a_1 a_2 \ldots a_n = G_r L_{n-r},$

where G_r denotes the g.c.d. of all the products of the a_i taken r at a time and L_{n-r} denotes the l.c.m. of all the products of the a_i taken $n - r$ at a time.

N.T./2 (1976/3).   Determine all integral solutions of

$$a^2 + b^2 + c^2 = a^2 b^2.$$

Solution.   We show, by considering the equation modulo 4 for all possibilities of a, b, c being even or odd, that it is necessary they all be even. We can also take them to all be non-negative. First note that for even and odd numbers, we have

$$(2n)^2 \equiv 0 \,(\mathrm{mod}\ 4) \qquad \text{and} \qquad (2n + 1)^2 \equiv 1 \,(\mathrm{mod}\ 4).$$

Case 1. a, b, c all odd. Then

$$a^2 + b^2 + c^2 \equiv 3 \,(\mathrm{mod}\ 4) \qquad \text{while} \quad a^2 b^2 \equiv 1 \,(\mathrm{mod}\ 4).$$

Cases 2. Two odd and one even. Then

$$a^2 + b^2 + c^2 \equiv 2 \,(\mathrm{mod}\ 4) \qquad \text{while} \quad a^2 b^2 \equiv 0 \text{ or } 1 \,(\mathrm{mod}\ 4).$$

Cases 3. Two even and one odd. Then

$$a^2 + b^2 + c^2 \equiv 1 \,(\mathrm{mod}\ 4) \qquad \text{while} \quad a^2 b^2 \equiv 0 \,(\mathrm{mod}\ 4).$$

Since the only possible solution is for a, b, c even, let $a = 2a_1$, $b = 2b_1$, and $c = 2c_1$. This leads to the equation

$$a_1^2 + b_1^2 + c_1^2 = 4a_1^2 b_1^2, \qquad \text{where } a_1 \leqslant a, \ b_1 \leqslant b, \ c_1 \leqslant c.$$

Now $4a_1^2 b_1^2 \equiv 0 \,(\mathrm{mod}\ 4)$, and each of a_1^2, b_1^2, c_1^2 is congruent to 0 or 1 (mod 4). Hence $a_1^2 \equiv b_1^2 \equiv c_1^2 \equiv 0 \,(\mathrm{mod}\ 4)$ and a_1, b_1, c_1 are even, say $a_1 = 2a_2$, $b_1 = 2b_2$, $c_1 = 2c_2$. This leads to the equation

$$16a_2^2 b_2^2 = a_2^2 + b_2^2 + c_2^2.$$

Again we can conclude that a_2, b_2, c_2 are all even, and the process leads to

$$64a_3^2 b_3^2 = a_3^2 + b_3^2 + c_3^2,$$

where $a = 8a_3$, $b = 8b_3$, $c = 8c_3$. If we continue the process, we conclude that a, b and c are divisible by as high a power of 2 as we want to specify, and hence the only solution of the equation is $a = b = c = 0$. This solution is an example of Fermat's method of infinite descent.

In a similar fashion, one can show that there are no non-trivial solutions of the Diophantine equations

(1) $$x^2 + y^2 + z^2 = x^2 y^2 z^2,$$

(2) $$x^2 + y^2 + z^2 = 2xyz,$$

(3) $$x^2 + y^2 + z^2 + w^2 = 2xyzw.$$

If we consider a generalization of (1), i.e.,

(4) $$x_1^r + x_2^r + \cdots + x_n^r = x_1^r x_2^r \cdots x_n^r, \qquad (2 < n < 2^r),$$

the infinite descent method is not as suitable as an inequality approach. Assuming that $0 < x_1 \leqslant x_2 \leqslant \cdots \leqslant x_n$, we have $nx_n^r \geqslant x_1^r x_2^r \cdots x_n^r$. This is possible only if $x_1 = x_2 = \cdots = x_{n-1} = 1$, and so there are no positive solutions of (4).

In [49; pp. 228–230], it is shown that only for $k = 1$ and 3 are there positive solutions to $x^2 + y^2 + z^2 = kxyz$, and an algorithm is given to generate all the solutions starting from the smallest one.

N.T./3. (1979 / 1). Determine all non-negative integral solutions $(n_1, n_2, \ldots, n_{14})$ if any, apart from permutations, of the Diophantine equation

$$n_1^4 + n_2^4 + \cdots + n_{14}^4 = 1{,}599.$$

Solution. We will show that there are no solutions. We will use the modular technique as in the previous solution. However, this time we use modulo 16. First note that $(2n)^4 \equiv 0 \pmod{16}$ and that $(2n + 1)^4 \equiv 8n(n + 1) + 1 \equiv 1 \pmod{16}$. Therefore

$$\sum_1^{14} n_i^4 \text{ can only be } \equiv 0, 1, 2, \ldots, 14 \pmod{16}, \text{ while } 1{,}599 \equiv 15 \pmod{16}.$$

This technique of proving the non-existence of integral solutions is very useful, but the rub is finding the appropriate modulus.

N.T./4. (1985/1). Determine whether or not there are any positive integral solutions of the simultaneous equations

$$x_1^2 + x_2^2 + \cdots + x_{1985}^2 = y^3,$$

$$x_1^3 + x_2^3 + \cdots + x_{1985}^3 = z^2$$

with distinct integers $x_1, x_2, \ldots, x_{1985}$.

Solution. More generally we will show that there are infinitely many integral solutions of the simultaneous equations

$$x_1^2 + x_2^2 + \cdots + x_n^2 = y^3,$$
$$x_1^3 + x_2^3 + \cdots + x_n^3 = z^2$$

for $n = 1, 2, 3, \ldots$.

Let

$$s = a_1^2 + a_2^2 + \cdots + a_n^2, \qquad t = a_1^3 + a_2^3 + \cdots + a_n^3,$$

where $a_1, a_2, \ldots, a_n$ is any set of positive integers. We now look for positive integers m and k such that $x_i = s^m t^k a_i$ will satisfy the given equations, i.e.,

$$x_1^2 + x_2^2 + \cdots + x_n^2 = s^{2m+1} t^{2k} = y^3,$$
$$x_1^3 + x_2^3 + \cdots + x_n^3 = s^{3m} t^{3k+1} = z^2.$$

Thus we need only satisfy

$$2m + 1 \equiv 2k \equiv 0 \ (\text{mod } 3) \quad \text{and} \quad 3m \equiv 3k + 1 \equiv 0 \ (\text{mod } 2).$$

These are satisfied for all $m \equiv 4 \ (\text{mod } 6)$ and $k \equiv 3 \ (\text{mod } 6)$.

The latter method can be extended to other systems of equations.

Exercise: Find infinitely many solutions of the simultaneous Diophantine equations

$$x_1^2 + x_2^2 + \cdots + x_n^2 = a^5,$$
$$x_1^3 + x_2^3 + \cdots + x_n^3 = b^2,$$
$$x_1^5 + x_2^5 + \cdots + x_n^5 = c^3.$$

N.T./5. (1978/3). An integer n will be called *good* if we can write

$$n = a_1 + a_2 + \cdots + a_k,$$

where $a_1, a_2, \ldots, a_k$ are positive integers (not necessarily distinct) satisfying

$$\frac{1}{a_1} + \frac{1}{a_2} + \cdots + \frac{1}{a_k} = 1.$$

Given the information that the integers 33 through 73 are good, prove that every integer ≥ 33 is good.

Solution. From a good integer n, we produce the two larger good integers $2n + 8$ and $2n + 9$ as follows: Let $(a_1, a_2, \ldots, a_k)$ be a partition of n which is good; then

$$\frac{1}{2a_1} + \frac{1}{2a_2} + \cdots + \frac{1}{2a_k} = \frac{1}{2}.$$

Since $\frac{1}{2} = \frac{1}{4} + \frac{1}{4} = \frac{1}{3} + \frac{1}{6}$, it follows that the two partitions

$$(4, 4, 2a_1, 2a_2, \ldots, 2a_k) \quad \text{and} \quad (3, 6, 2a_1, 2a_2, \ldots, 2a_k)$$

also have the property that the sum of the reciprocals is 1. These are partitions of the integers $2n + 8$ and $2n + 9$, respectively. Therefore,

(1)   if n is good, so also are $2n + 8$ and $2n + 9$.

So "33 is good" implies that 74, 75 are good. We use the hypothesis to fill the gap between $n = 33$ and $2n + 8 = 74$: Let S_n denote the statement "all the integers $n, n + 1, \ldots, 2n + 7$ are good". We begin an induction with the given information that S_{33} is valid. By (1) we conclude that $S_n \Rightarrow S_{n+1}$. Hence by induction S_n is valid for all $n \geq 33$, giving the desired result.

N.T./6 (1973/5). Show that the cube roots of three distinct prime numbers cannot be three terms (not necessarily consecutive) of an arithmetic progression.

Solution. Our proof is indirect, so we assume they can be. Hence,

$$\sqrt[3]{p} = a, \qquad \sqrt[3]{q} = a + md, \qquad \sqrt[3]{r} = a + nd$$

where p, q, r are the distinct prime numbers and m, n are integers. Eliminating a and d, we obtain

$$\left(\sqrt[3]{q} - \sqrt[3]{p}\right) \Big/ \left(\sqrt[3]{r} - \sqrt[3]{p}\right) = m/n$$

or

(1) $$m\sqrt[3]{r} - n\sqrt[3]{q} = (m - n)\sqrt[3]{p}.$$

Cubing (1), we obtain

(2) $$m^3 r - n^3 q - 3mn\sqrt[3]{rq}\left(m\sqrt[3]{r} - n\sqrt[3]{q}\right) = (m - n)^3 p.$$

Now using (1) to replace $m\sqrt[3]{r} - n\sqrt[3]{q}$, we get

$$3mn(m - n)\sqrt[3]{prq} = m^3 r - n^3 q - (m - n)^3 p.$$

This gives the contradiction, since $\sqrt[3]{pqr}$ is irrational.

Note that p, q, r need not be distinct prime numbers. They can be arbitrary integers with the restriction that none, nor their product is a perfect cube. I had conjectured that the result could be extended to the n-th roots of 3 distinct prime numbers. This was subsequently proved by Bonnie Stewart in a private communication. However, the proof involves considerably more "algebra". Much later this problem resurfaced as a proposed problem (Amer. Math. Monthly (1977) 202) by R. Hall in the form "Let a, b, c be distinct positive integers, at least two of which are prime. Show that $\sqrt[n]{a}$, $\sqrt[n]{b}$, and $\sqrt[n]{c}$ cannot be distinct terms of an arithmetic progression."

Exercise: Show that the cube roots of three distinct prime numbers cannot be three terms (not necessarily consecutive) of a geometric, nor of a harmonic progression.

N.T./7. (1984/2). The geometric mean of any set of m non-negative numbers is the m-th root of their product.

(i) For which positive integers n is there a finite set S_n of n distinct positive integers such that the geometric mean of any subset of S_n is an integer?

(ii) Is there an infinite set S of distinct positive integers such that the geometric mean of any finite subset of S is an integer?

Solution. (i) S_n exists for every positive integer n. Just let S_n consist of any n distinct positive integers each of which is a perfect $(n!)$ power.

(ii) We show that there are no such sets by an indirect proof. Assume to the contrary that there is such a set S. Let a and b be two elements of S. By the unique factorization theorem, each has a finite number of prime factors, so b/a can be a rational n-th power for only finitely many n. Hence there exists an integer p such that $\sqrt[p]{b/a}$ is irrational. Now take any other integers $c_2, c_3, \ldots, c_p$ from the infinite set S. By hypothesis, the geometric means of $(b, c_2, c_3, \ldots, c_p)$ and $(a, c_2, c_3, \ldots, c_p)$ are both integers. This is a contradiction since the ratio of the two means is the irrational number $\sqrt[p]{b/a}$.

N.T./8. (1986/3). What is the smallest integer n, greater than one, for which the root-mean-square of the first n positive integers is an integer?

Note. The root-mean square of n numbers $a_1, a_2, \ldots, a_n$ is defined to be

$$\left[\frac{a_1^2 + a_2^2 + \cdots + a_n^2}{n} \right]^{1/2}.$$

Solution. It is a known result (which can be established by induction or otherwise) that

$$1^2 + 2^2 + \cdots + n^2 = \frac{n(n+1)(2n+1)}{6}.$$

Consequently we want the least $n > 1$ satisfying the Diophantine equation $(n + 1)(2n + 1) = 6m^2$. Now 6 divides $(n + 1)(2n + 1)$ if and only if $n \equiv 5$ or 1 (mod 6). In symbols:

$$6 | (n + 1)(2n + 1) \ iff \ n \equiv 5 \ or \ 1 \ (mod \ 6).$$

Case 1. $n = 6k + 5$.

Here $m^2 = (k + 1)(12k + 11)$. Since the two factors are relatively prime, each must be a square, i.e., $k + 1 = a^2$ and $12k + 11 = b^2$. Then $12a^2 = b^2 + 1$. As in N.T./2 and 3, the latter equation can be shown to be impossible modulo 4; here $12a^2 \equiv 0$ (mod 4) whereas $b^2 + 1 \equiv 1$ or 2 (mod 4).

Case 2. $n = 6k + 1$.

Here $m^2 = (3k + 1)(4k + 1)$. Again since the factors are relatively prime, we must have $3k + 1 = a^2$, $4k + 1 = b^2$. Then $(2a - 1)(2a + 1) = 3b^2$. Any possible prime factor of the right hand side except 3 must appear to an even power. We now try successive values of a starting with 1 and such that neither $2a \pm 1$ is a prime other than 3; $a = 1$ leads to $b = 1$ and to $n = 1$ which is not > 1. The next smallest suitable value of a is 13 so that $n = 337$ is the required result.

N.T./9. (1975/1). (a) Prove that
$$[5x] + [5y] \geqslant [3x + y] + [3y + x],$$
where $x, y \geqslant 0$ and $[u]$ denotes the greatest integer $\leqslant u$ (e.g., $[\sqrt{2}] = 1$).

(b) Using (a) or otherwise, prove that
$$\frac{(5m)!(5n)!}{m!n!(3m + n)!(3n + m)!}$$
is integral for all positive integral m and n.

Solution. (a) Let $x = x_1 + u$ and $y = y_1 + v$, where x_1 and y_1 are nonnegative integers and $0 \leqslant u < 1$, $0 \leqslant v < 1$. The inequality to be proved then reduces to

(1) $\qquad x_1 + y_1 + [5u] + [5v] \geqslant [3u + v] + [3v + u]$.

We prove that

(2) $\qquad [5u] + [5v] \geqslant [3u + v] + [3v + u]$,

which implies (1). In view of the symmetric roles of u and v we may assume $u \geqslant v$, which gives the inequality $[5u] \geqslant [3u + v]$. If $u \leqslant 2v$, we also have $[5v] \geqslant [3v + u]$, and so (2) is established in this case.

Finally we prove that (2) holds if $u > 2v$. Let $5u = a + b$ and $5v = c + d$, where a and c are nonnegative integers, and $0 \leqslant b < 1$, $0 \leqslant d < 1$. Then (2) can be rewritten as

(3) $\qquad a + c \geqslant [(3a + c + 3b + d)/5] + [(3c + a + 3d + b)/5]$.

Now $1 > u > 2v$, so that $5 > 5u > 10v$ or $5 > a + b > 2c + 2d$. The first inequality here gives $5 > a$ or $4 \geqslant a$. The second inequality gives $a \geqslant 2c$, because $a < 2c$ would imply that $a \leqslant 2c - 1$, $a + 1 - 2c \leqslant 0$ and $a + b - 2c < 0$. Thus we have $4 \geqslant a \geqslant 2c$, and hence the only cases we need to consider are these:

a	4	4	4	3	3	2	2	1	0
c	2	1	0	1	0	1	0	0	0

It is easy to verify the inequality (3) in these nine cases, because $3b + d < 4$ and $3d + b < 4$.

(b) Since the highest power of a prime p dividing $m!$ is

$$[m/p] + [m/p^2] + [m/p^3] + \cdots,$$

it suffices to show that

$$(4) \qquad \left[\frac{5m}{r}\right] + \left[\frac{5n}{r}\right] \geqslant \left[\frac{m}{r}\right] + \left[\frac{n}{r}\right] + \left[\frac{3m + n}{r}\right] + \left[\frac{3n + m}{r}\right]$$

for arbitrary integers $r \geqslant 2$. Letting $m = rm_1 + x$, $n = rn_1 + y$, where $0 \leqslant x < r$, $0 \leqslant y < r$, and r, m_1, n_1 are integers, (4) becomes

$$[5x/r] + [5y/r] \geqslant [(3x + y)/r] + [(3y + x)/r],$$

which follows from part (a).

Exercises: Show that the following expressions are all integral:

(i)
$$\frac{(3m)!(3n)!}{m!n!(m + n)!(m + n)!},$$

(ii)
$$\frac{(4m)!(4n)!}{m!n!(2m + n)!(2n + m)!},$$

(iii)
$$\frac{(mnr)!}{m!n!^m r!^{mn}},$$

(iv)
$$\frac{(m - 1)!\delta}{m_1!m_2! \cdots m_r!},$$

where $m = m_1 + m_2 + \cdots + m_r$ and δ is the g.c.d. of $m_1, m_2, \ldots, m_r$.

For a graphical solution of these kinds of problems, see solution by Greg Patruno, (a former USAMO winner) in *Amer. Math. Monthly*, 94 (1987) pp. 1012–1014.

N.T./10. (1982/4). Prove that there exists a positive integer k such that $k \cdot 2^n + 1$ is composite for every positive integer n.

Solution. We first consider the less ambitious problem of finding a positive integer k such that $k \cdot 2^n + 1$ is composite for all positive integers n in an infinite arithmetic series.

Let b be an integer > 1, and let p be a prime divisor of $2^b - 1$, so that

$$(1) \qquad\qquad 2^b \equiv 1 \;(\text{mod } p).$$

Let a be any integer such that $0 \leqslant a < b$, and let k be an integer $> p$ such that

$$k \equiv -2^{b-a} \;(\text{mod } p).$$

If n is a positive integer satisfying

$$(2) \qquad\qquad n \equiv a \;(\text{mod } b),$$

i.e., $n = a + bm$ for some integer $m \geqslant 0$, then by (1)

$$k2^n \equiv -2^{b-a}2^{a+bm} \equiv -1 \;(\text{mod } p).$$

Thus $k \cdot 2^n + 1$ is divisible by p, and since it is greater than p, it is indeed composite for all n satisfying (2).

The desired result will now follow if we can construct a *finite* set of triples (p_j, a_j, b_j) with the following properties: The p_j are *distinct* primes. The b_j are positive integers such that for each j,

$$(1)_j \qquad\qquad 2^{b_j} \equiv 1 \;(\text{mod } p_j).$$

The a_j are integers, $0 \leqslant a_j < b_j$, such that every integer n satisfies at least one of the congruences

$$(2)_j \qquad\qquad n \equiv a_j \;(\text{mod } b_j).$$

Assuming that the above construction has been accomplished, we use the Chinese Remainder Theorem to obtain a positive integer k, greater than each p_j, such that for each j,

$$k \equiv -2^{b_j - a_j} \;(\text{mod } p_j).$$

Then $k \cdot 2^n + 1$ would be composite for every n.

The only remaining step is to carry out the above construction. This can be done in a number of ways and we shall give two different approaches.

Construction 1. To verify that every integer n satisfies at least one of the congruences $(2)_j$, it suffices to test every residue class modulo b, where b is the least common multiple of the b_j. Thus it is reasonable to choose an integer b with many divisors. An appropriate choice is $b = 24$, whose divisors (> 1) are $2, 3, 4, 6, 8, 12,$ and 24. By condition $(1)_j$, $2^{b_j} - 1$ must be divisible by p_j. Selecting b_j from the list above, we note that

$$2^2 - 1 = 3, \quad 2^3 - 1 = 7, \quad 2^4 - 1 = 3 \cdot 5, \quad 2^6 - 1 = 3^2 \cdot 7,$$
$$2^8 - 1 = 3 \cdot 5 \cdot 17, \quad 2^{12} - 1 = 3^2 \cdot 5 \cdot 7 \cdot 13,$$
$$2^{24} - 1 = 3^2 \cdot 5 \cdot 7 \cdot 13 \cdot 17 \cdot 241.$$

Since the primes p_j have to be different, 6 *may not* be chosen as one of the b_j if both 2 and 3 are to be among them. The triples may now be chosen as follows:

b_j	2	3	4	8	12	24
a_j	0	0	1	3	7	23
p_j	3	7	5	17	13	241

The a_j are picked after simple experimentation and it is easy to verify that every integer between 1 and 24 inclusive satisfies at least one of the congruences $(2)_j$.

Construction 2. Take $b = 2^h$ as the least common multiple of the b_j and choose

$$b_j = 2^j, \qquad a_j = 2^{j-1} - 1 \qquad \text{for } 1 \leqslant j \leqslant h$$

and

$$b_{h+1} = 2^h, \qquad a_{h+1} = 2^h - 1.$$

We claim that each integer between 1 and 2^h inclusive satisfies exactly one of the congruences $(2)_j$. Clearly each integer n satisfies at most one of these congruences. Exactly 2^{h-j} of these integers satisfy the congruence $(2)_j$ for $1 \leqslant j \leqslant h$ while exactly one of them satisfies $(2)_{h+1}$. Since

$$2^{h-1} + 2^{h-2} + \cdots + 1 + 1 = 2^h,$$

no integer between 1 and 2^h is left out.

That two of the b_j are equal poses no problem since it is the p_j that have to be distinct. By conditions $(1)_j$, the p_j have to be divisors of $2^{b_j} - 1 = 2^{2^j} - 1$. We claim that they can be chosen to be distinct.

We proceed inductively. Since $2^{2^1} - 1 = 3$, we must take $p_1 = 3$. Suppose $p_1, p_2, \ldots, p_m$ have been chosen. Write

$$2^{2^{m+1}} - 1 = (2^{2^m} - 1)(2^{2^m} + 1).$$

We choose p_{m+1} to be one of the primes dividing $2^{2^m} + 1$. Since for $1 \leqslant j \leqslant m$, $2^{2^j} - 1$ divides $2^{2^m} - 1$, it is relatively prime to $2^{2^m} + 1$. This shows that none of the p_j, $1 \leqslant j \leqslant m$, equals p_{m+1}.

Note that p_{m+1} is uniquely defined as long as $2^{2^m} + 1$ is a prime or a prime power. It turns out, as observed by Mersenne, that $2^{2^m} + 1$ is prime for $m = 0, 1, 2, 3, 4$, while Euler had determined that $2^{2^5} + 1$ is the product of the two primes 641 and 6,700,417. It follows that we can take $h = 6$ and choose the triples as follows:

b_j	2	4	8	16	32	64	64
a_j	0	1	3	7	15	31	63
p_j	3	5	17	257	65,537	641	6,700417.

Other possible sets of triples along the lines of construction 1 are given by the following two exercises:

Exercise 1. Choose $b = 36$ with

b_j	2	3	4	9	12	18	36
a_j	0	0	1	2	7	5	35
p_j	3	7	5	73	13	19	37

Verify $(1)_j$ and show that every integer n, $1 \leqslant n \leqslant 36$, satisfies at least one of the congruences $(2)_j$.

Exercise 2. Choose $b = 60$ with

b_j	2	3	4	5	10	12	15	20	30
a_j	0	0	1	0	3	11	7	19	1
p_j	3	7	5	31	11	13	151	41	331

Verify $(1)_j$ and show that every integer n, $1 \leqslant n \leqslant 60$, satisfies at least one of the congruences $(2)_j$.

Comment. Both constructions 1 and 2 go back to Sierpinski[†],[‡] who noted the following related results:

(i) From construction 1, it can be shown that there exist infinitely many odd integers k such that all the numbers $2^n + k$, $n = 1, 2, \ldots$, are composite.

(ii) Erdös showed that there exist infinitely many odd natural numbers k such that $2^n + k$ $(n = 1, 2, \ldots)$ is divisible by one of $3, 5, 13, 17, 241$. Schinzel proved a similar statement with the above set of primes replaced by the set of primes < 100.

Schinzel also showed the following: Suppose we have an integer P and an odd integer k, such that for $n = 1, 2, \ldots$, $2^n + k$ is divisible by a prime factor of P. Then $k2^n + 1$ is also divisible by a prime factor of P for $n = 1, 2, \ldots$.

Schinzel's proof is as follows: By our hypothesis, $2^{n(\phi(P)-1)} + k$ is divisible by a prime divisor p of P. By the Fermat-Euler theorem (see Glossary), $2^{n\phi(P)} \equiv 1 \pmod{P}$, and hence

(1)
$$2^{n\phi(P)} \equiv 1 \pmod{p}.$$

We have

$$2^{n\phi(P)-n} \equiv -k \pmod{p}.$$

If we multiply this by 2^n and use (1), we get $k2^n \equiv -1 \pmod{p}$ or $p \mid k2^n + 1$.

The smallest k for which $k2^n + 1$ is composite for all n is not known. However, several bounds on it are reported by Guy.[§]

N.T./11. (1983/5). Consider an open interval of length $1/n$ on the real number line, where n is a positive integer. Prove that the number of

[†] W. Sierpinski, *250 Problems in Elementary Number Theory*, American Elsevier, N.Y., 1970, p. 64, #119.

[‡] W. Sierpinski, Sur un problème concernant les nombres $k \cdot 2^n + 1$, Elem. Math. 16 (1960) 73–74, and corrigendum 17 (1962) 85.

[§] R. Guy, *Unsolved Problems in Number Theory*, Springer-Verlag, N.Y., 1981, pp. 42–43.

irreducible fractions p/q, with $1 \leqslant q \leqslant n$, contained in the given interval is at most $(n + 1)/2$.

Solution. We divide all the rational points in $(\alpha, \alpha + \frac{1}{n})$ into two sets: $\{u_i/v_i\}$, $i = 1, 2, \ldots, r$, with denominators v_i between 1 and $n/2$ inclusive, and $\{x_i/y_i\}$, $i = 1, 2, \ldots, s$, with denominators $y_i > n/2$ and $\leqslant n$, where all these fractions are in reduced form. For every v_i there exist integers c_i, such that $n/2 \leqslant c_i v_i \leqslant n$. Define $y_{s+i} = c_i v_i$ and $x_{s+i} = c_i u_i$. No two elements of $\{y_i: 1 \leqslant i \leqslant r + s\}$ are equal, for $y_j = y_k$ implies that $|x_j/y_j - x_k/y_k| \geqslant 1/y_j \geqslant 1/n$. This contradicts that the open interval is of length $1/n$. Hence the number of distinct rational points is $r + s \leqslant n - [n/2] \leqslant (n + 1)/2$.

Note that this maximum number can be attained on the interval $(\varepsilon + 1/n, \varepsilon + 2/n)$ with $0 < \varepsilon < 1/(n - 1) - 1/n$. The rational numbers in this interval are $1/(n - 1)$, $1/(n - 2), \ldots, 1/[(n + 1)/2]$ and $2/n$. The last two rational numbers are distinct if n is odd and identical if n is even.

Another proof can be based on the following known result: If $\{a_i: 1 \leqslant i \leqslant n + 1\}$ is a subset of $\{1, 2, \ldots, 2n\}$, then there exist a_j and a_k such that a_j divides a_k. (To prove this take the set $\{b_i: 1 \leqslant i \leqslant n + 1\}$ where b_i is the largest odd integer dividing a_i and use the pigeon hole principle.) In the present problem, suppose on the contrary we have at least $k + 1$ fractions where $k = [(n + 1)/2]$. Let $\{q_i: 1 \leqslant i \leqslant k + 1\}$ be the set of the denominators. This is a subset of $\{1, 2, \ldots, 2k\}$. By the previous result, q_j divides q_k for some j and k, say $q_k = \alpha q_j$. Hence

$$|p_j/q_j - p_k/q_k| = |\alpha p_j - p_k|/q_k \geqslant 1/n,$$

contradicting the hypothesis that the open interval is of length $1/n$.

N.T./12. (1986/1).

(a) Do there exist 14 consecutive positive integers each of which is divisible by one or more primes p from the interval $2 \leqslant p \leqslant 11$?

(b) Do there exist 21 consecutive positive integers each of which is divisible by one or more primes p from the interval $2 \leqslant p \leqslant 13$?

Solution. (a) The answer is no. First note that we can ignore the 7 even numbers in any set of 14 consecutive integers since they are divisible by 2. We now show that of the remaining seven integers, at least one is not divisible by 3, 5, 7, or 11. Of these seven odd integers, at most three are divisible by 3, at most two are divisible by 5, at most one is divisible by 7, and at most one is divisible by 11. It follows that each of the seven odd integers can be divisible by 3, 5, 7, or 11 only if each of these primes divides its maximum number of terms and there are no overlaps (i.e., no term is divisible by a product of these primes). To see this is impossible, first note that in order to get three of the terms divisible by 3, they must be the first term, the middle term, and the last term of the seven odd numbers. Then since the numerical difference between any of the remaining terms is less than 10, only one of these can be divisible by 5.

(b) The answer is yes. First note that if n is divisible by $2 \cdot 3 \cdot 5 \cdot 7 = 210$, then of the 21 consecutive numbers $n - 10, n - 9, \ldots, n + 10$, all but $n - 1$ and $n + 1$ are divisible by a prime ≤ 7. We now determine n such that $n - 1$ and $n + 1$ are divisible by 11 or 13. By the Chinese Remainder Theorem, positive integers n exist satisfying the simultaneous congruences

$$n \equiv 0 \ (\mathrm{mod}\ 210), \qquad n \equiv 1 \ (\mathrm{mod}\ 11), \qquad n \equiv -1 \ (\mathrm{mod}\ 13).$$

For any such n, we get the desired result. In particular, the smallest such n is 9450 and the 21 consecutive numbers with the desired divisibility properties are $9440, 9441, \ldots, 9460$. Also, it turns out, this is the smallest set of consecutive positive integers with the given property.

N.T./13. (1986/5). By a *partition π of an integer $n \geq 1$*, we mean here a representation of n as a sum of one or more positive integers where the summands must be put in nondecreasing order. (E.g., if $n = 4$, then the partitions π are $1 + 1 + 1 + 1$, $1 + 1 + 2$, $1 + 3$, $2 + 2$, and 4.)

For any partition π, define $A(\pi)$ to be the number of 1's which appear in π, and define $B(\pi)$ to be the number of distinct integers which appear in π. (E.g., if $n = 13$ and π is the partition $1 + 1 + 2 + 2 + 2 + 5$, then $A(\pi) = 2$ and $B(\pi) = 3$.)

Prove that, for any fixed n, the sum of $A(\pi)$ over all partitions π of n is equal to the sum of $B(\pi)$ over all partitions π of n.

Solution. Let $P(n)$ denote the number of partitions of n; define $P(0) = 1$. The number of partitions of n in which the number m occurs at least once is then $P(n - m)$. Using this result, we first prove by induction that

$$(1) \qquad \sum A(\pi) = P(n - 1) + P(n - 2) + \cdots + P(1) + P(0),$$

where the summations here and subsequently are over all partitions π of n. The result is clearly valid for $n = 1$. Now assume that (1) is valid for all $k < n$ where $n > 1$. The number of partitions of n in which one or more 1's appear is $P(n - 1)$ and these are of the form $1 + \{\text{partition of } (n - 1)\}$. Of the $P(n)$ partitions, there are $P(n - 1)$ starting with a 1 and, by the inductive hypothesis, the partitions of $(n - 1)$ contribute another $[P(n - 2) + P(n - 3) + \cdots + P(0)]$ 1's. Thus (1) follows by induction.

To complete our solution, we now show also that

$$(2) \qquad \sum B(\pi) = P(n - 1) + P(n - 2) + \cdots + P(1) + P(0).$$

Consider a $P(n)$ by n rectangular array with the rows indexed by the partitions π of n and the columns indexed by the numbers $1, 2, \ldots, n$. A checkmark is placed in row π and column m if the number m appears in that partition. Thus $\Sigma B(\pi) = $ the total number of check marks in the array. On the other hand, the number of checkmarks in the column m of the array is the number of partitions in which an m occurs and this is $P(n - m)$. Thus adding up the checkmarks in each column gives (2). (This result is attributed to Richard Stanley.)

Our next solution is via generating functions† and is due to Loren Larson and Bruce Hanson.

Let

$$P_k(x) = 1 + x^k + x^{2k} + x^{3k} + \cdots,$$

$$Q(x) = x + 2x^2 + 3x^3 + \cdots,$$

$$R(x) = Q(x) \prod_{k=2}^{\infty} P_k(x).$$

After considering how the coefficients of $R(x)$ are formed, we see that the coefficient of x^n equals $\Sigma A(\pi)$. Also

$$Q(x) = x + 2x^2 + 3x^3 + \cdots$$

$$= (x + x^2 + x^3 + \cdots)(1 + x + x^2 + \cdots)$$

$$= (x + x^2 + x^3 + \cdots)P_1(x),$$

so that

$$R(x) = (x + x^2 + x^3 + \cdots) \prod_{k=1}^{\infty} P_k(x).$$

Now letting

$$\prod_{k=1}^{\infty} P_k(x) = c_0 + c_1 x + c_2 x^2 + \cdots,$$

and examining how the c_i are formed, we see that the coefficient c_{n-k} is the number of partitions of n which contain the integer k for $1 \leq k \leq n$. Thus the number of partitions of n in which k occurs, summed for $k = 1, 2, 3, \ldots, n$, is $c_{n-1} + c_{n-2} + \cdots + c_0$, and this is also the coefficient of x^n in $R(x) = (x + x^2 + x^3 + \cdots) \prod_{k=1}^{\infty} P_k(x)$. But observe that the number of partitions π of n in which k occurs, summed for $k = 1, 2, 3, \ldots, n$, is precisely the sum of $B(\pi)$ over all partitions π of n. This completes the proof.

†For other applications of generating functions to other partition functions, see G.E. Andrews, *The Theory of Partition Functions*, Addison-Wesley, Reading, 1976.

Plane Geometry

P.G./1. (1981/1). The measure of a given angle is $180°/n$ where n is a positive integer not divisible by 3. Prove that the angle can be trisected by Euclidean means (straight edge and compasses).

Solution. Since 3 and n are relatively prime, there are integers r, s such that $3r + ns = 1$. Multiplying the latter equation by $60°/n$, we get

$$\frac{180°r}{n} + 60°s = \frac{60°}{n}.$$

Since the angle $180°/n$ is given, and since we can construct a $60°$ angle, we then can construct the angles $180°r/n$ and $60°s$ and thus the angle $60°/n$.

More generally we can m-sect a given angle $180°/n$ provided m is relatively prime to n and the angle $180°/m$ is constructible. This follows as before. First there are integers r, s such that $mr + ns = 1$. Multiplying by $180°/mn$, we get

$$\frac{180°r}{n} + \frac{180°s}{m} = \frac{180°}{mn}.$$

We can construct an angle of $180°/m$ if a regular m-gon can be constructed. It was shown by Gauss that a regular polygon having a prime number p of sides can be constructed by Euclidean means *if and only if* p is of the form $p = 2^{2^n} + 1$. For $n = 0, 1, 2, 3, 4$, we find that p are the prime numbers $3, 5, 17, 257, 65537$. Many Euclidean constructions of the regular polygon of 17 sides (the regular heptadecagon) have been given. In particular, see [60; pp. 173–176] or R. C. Yates, Geometric Tools, Educational Publishers, St. Louis, 1949, p. 30. A manuscript deposited in Göttingen contains a very complicated construction for a regular polygon of 257 sides.

P.G./2. (1978/ 32). *ABCD* and *A'B'C'D'* are square maps of the same region, drawn to different scales and superimposed as shown in the figure. Prove that there is only one point O on the small map which lies directly over point O' of the large map such that O and O' each represent the same place of the country. Also, give a Euclidean construction (straight edge and compasses) for O.

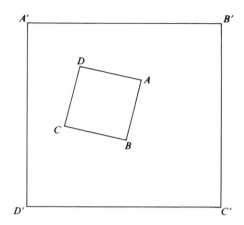

Solution. The problem is solved by means of the theorem that a contraction mapping on the closed square has a unique fixed point. This theorem is a special case of the Brouwer Fixed Point Theorem [20; pp. 278–79], but may be proved directly by iterating the contraction mapping. In our problem, this would give a sequence of ever smaller similarly oriented squares within the previous set of squares. We get a sequence of nested regions whose diameters tend to 0. The *limit point* or *intersection* of this set of nested regions is the unique fixed point.

Since the two given squares are directly similar, there is a unique spiral similarity that carries one square into the other.† We now give two constructions for the fixed point. The first one uses circles while the second one uses only a straight edge.

Let AB intersect $A'B'$ at point P and construct circles through A, P, A' and B, P, B'. The circles intersect at the desired fixed point O. To see this, first note that the angles labeled α in the figure are equal since both are measured by half the arc PO. Also the angles labeled β are equal since both are supplementary to $\angle PAO$. Hence, $\triangle OAB \sim \triangle OA'B'$ and so a rotation about O through $\angle AOA'$ places the square $ABCD$ with its sides parallel to $A'B'C'D'$, and a dilatation with ratio $A'B'/AB$ makes them coincide.

For the straight-edge construction, first find the intersections of AB with $A'B', BC$ with $B'C', CD$ with $C'D'$, and DA with $D'A'$ and denote

---

†See H. S. M. Coxeter and S. L. Greitzer, *Geometry Revisited*, NML vol. 19, p. 85.

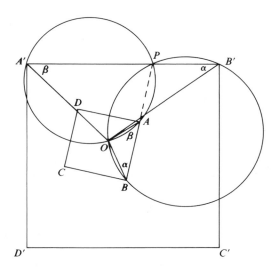

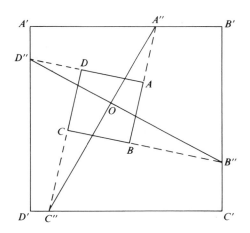

them by A'', B'', C'', D'', respectively. Then the fixed point O (the center of similarity), is the intersection of $A''C''$ with $B''D''$. This follows since the figure of O and the two lines AB and CD is similar to the figure of O and the two lines $A'B'$ and $C'D'$. Thus the dilatation from O that takes line AB to CD also takes line $A'B'$ to $C'D'$, and A'' to C''. Consequently, O is collinear with A'' and C''. Similarly, O is collinear with B'' and D''. [85; pp. 72–74].

P.G./3. (1977/2).   *ABC* and *A'B'C'* are two triangles in the same plane such that the lines *AA'*, *BB'*, *CC'* are mutually parallel. Let [*ABC*] denote the area of triangle *ABC* with an appropriate ± sign, etc.; prove that

$$3([ABC] + [A'B'C']) = [AB'C'] + [BC'A'] + [CA'B']$$
$$+ [A'BC] + [B'CA] + [C'AB].$$

First Solution.   We remind the reader that the sign of [*ABC*] is defined as follows: We first designate a sense of rotation in the plane of the triangle as the positive sense of rotation. Then [*ABC*] will be positive if going around the triangle from *A* to *B* to *C* to *A* is a rotation in the designated positive sense, and it will be negative if it is a rotation in the opposite sense. Thinking of the plane as one side of opaque paper, one takes the counter-clockwise sense of rotation as positive. (On cellophane a given sense of rotation is clockwise or counterclockwise, depending on which side we view it from.)

We start with the following identity, which could be proved algebraically but is evident from the figure. If *O* is any point in the plane (it could be outside △*ABC*) then

(1) $$[ABC] = [OAB] + [OBC] + [OCA].$$

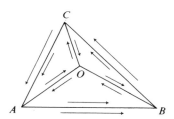

Thus

(2) $$[ABC] = [A'BC] + [A'CA] + [A'AB]$$

(2') $$[A'B'C'] = [AB'C'] + [AC'A'] + [AA'B'].$$

By assumption *CC'* ∥ *AA'*. Hence [*A'CA*] and [*AC'A'*] have the same absolute value, and since we go around them in opposite senses, their sum is 0. If we add (2) and (2') the other pair of terms containing *AA'* also cancels and we get

$$[ABC] + [A'B'C'] = [A'BC] + [AB'C'].$$

We get two similar equations by cyclically permuting A, B, C. Adding the three equations we get the required result.

Second solution.   Our second solution involves more calculations. Nevertheless, it gives some application of determinants which can be quite useful in other problems.

The signed area of a triangle whose vertices have rectangular coordinates (x_1, y_1), (x_2, y_2), and (x_3, y_3) is given by

$$F = \frac{1}{2}\begin{vmatrix} x_1 & y_1 & 1 \\ x_2 & y_2 & 1 \\ x_3 & y_3 & 1 \end{vmatrix}.$$

Secondly, since a determinant is a linear function of its columns,

$$(1)\qquad \begin{vmatrix} x_1 + x_1' & y_1 & 1 \\ x_2 + x_2' & y_2 & 1 \\ x_3 + x_3' & y_3 & 1 \end{vmatrix} = \begin{vmatrix} x_1 & y_1 & 1 \\ x_2 & y_2 & 1 \\ x_3 & y_3 & 1 \end{vmatrix} + \begin{vmatrix} x_1' & y_1 & 1 \\ x_2' & y_2 & 1 \\ x_3' & y_3 & 1 \end{vmatrix}.$$

We now coordinatize the vertices of the two given triangles as in the following figure:

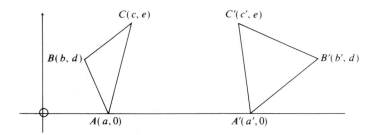

Expressing the area of the 8 specified triangles in determinant form, we have to show that

$$3\begin{vmatrix} a & 0 & 1 \\ b & d & 1 \\ c & e & 1 \end{vmatrix} + 3\begin{vmatrix} a' & 0 & 1 \\ b' & d & 1 \\ c' & e & 1 \end{vmatrix}$$

$$= \begin{vmatrix} a & 0 & 1 \\ b' & d & 1 \\ c & e & 1 \end{vmatrix} + \begin{vmatrix} a' & 0 & 1 \\ b & d & 1 \\ c & e & 1 \end{vmatrix} + \begin{vmatrix} a' & 0 & 1 \\ b & d & 1 \\ c' & e & 1 \end{vmatrix}$$

$$+ \begin{vmatrix} a & 0 & 1 \\ b' & d & 1 \\ c' & e & 1 \end{vmatrix} + \begin{vmatrix} a' & 0 & 1 \\ b' & d & 1 \\ c & e & 1 \end{vmatrix} + \begin{vmatrix} a & 0 & 1 \\ b & d & 1 \\ c' & e & 1 \end{vmatrix}.$$

Using property (1) on both sides of this equation, we obtain the equivalent equation

$$3\begin{vmatrix} a + a' & 0 & 1 \\ b + b' & d & 1 \\ c + c' & e & 1 \end{vmatrix} = \begin{vmatrix} 3(a + a') & 0 & 1 \\ 3(b + b') & d & 1 \\ 3(c + c') & e & 1 \end{vmatrix},$$

which is obviously true.

Note that the result here can be extended to tetrahedra and more generally to simplexes.

P.G./4, (1976/2). If A and B are fixed points on a given circle and XY is a variable diameter of the same circle, determine the locus of the point of intersection of lines AX and BY. You may assume that AB is not a diameter.

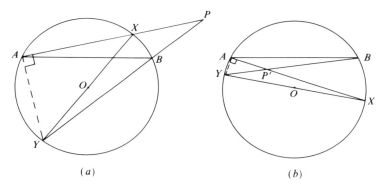

(a)   (b)

Solution. The figures above show two positions of the moving diameter XY. In either case, $\angle XAY = 90°$ and also $\angle AYB$ is constant. Thus in Figure (a), $\angle APB$ which is the complement of $\angle AYB$ is also constant. Similarly, in Figure (b), $\angle AP'Y$ is constant since it is the complement of $\angle AYB$. Since $\angle AP'B$ is the supplement of $\angle AP'Y$, it is also constant. In Figure (a), the locus of points P such that $\angle APB$ is constant is a circular arc on chord AB.

In Figure (b), the triangle with fixed base AB and constant $\angle AP'B$ has its vertex P' on a circular arc on chord AB. Since $\triangle APY$ of diagram (a) is similar to $\triangle AP'Y$ in diagram (b), angles APB and $AP'B$ are supplements. Thus P and P' lie on the same circle. The radius of this circle, by the extended Law of Sines, is $r = AB/2\sin P$ (see NML, vol. 19, pp. 1-2).

P.G./5. (1986/4). Two distinct circles K_1 and K_2 are drawn in the plane. They intersect at points A and B, where AB is a diameter of K_1. A point P on K_2 and inside K_1 is also given.

Using only a "T-square" (i.e., an instrument which can produce the straight line joining two points and the perpendicular to a line through a point on or off the line), find a construction for two points C and D on K_1 such that CD is perpendicular to AB and CPD is a right angle.

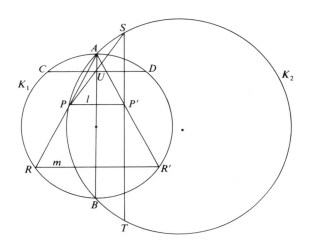

Solution. We first construct the mirror image P' of P with respect to AB; see figure. We draw line AP and denote its intersection with circle K_1 by R. Then we draw perpendicular lines l and m to AB from P and R, respectively. Line m intersects K_1 again at a point R'. P' is also the intersection of line AR' with l.

Through P' we now draw a line perpendicular to l which meets circle K_2 in points S and T. Draw line SP meeting AB in point U. The perpendicular to AB through U gives the desired chord CD. [If we also draw line TP meeting AB in a point U', we obtain a second pair of points C', D' that satisfy the conditions of the problem.]

PROOF: By the power of a point theorem, $SU \cdot UP = AU \cdot UB = CU \cdot UD$. Since AB bisects segment PP', it also bisects SP, so $SU = UP$; also $CU = UD$. It follows that $SU = CU$. Hence, U is the center of a circle passing through C, D, P, so that CPD is a right angle.

P.G./6. (1972/5). A given convex pentagon $ABCDE$ has the property that the area of each of the five triangles ABC, BCD, CDE, DEA, and EAB is unity. Show that all pentagons with the above property have the same area, and calculate that area. Show, furthermore, that there are infinitely many non-congruent pentagons having the above area property.

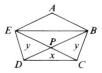

Solution. Let $[ABC]$ denote the area of triangle ABC, etc. Since $[EDC] = [BDC] = 1$, (see figure) both of these triangles have equal altitudes to side CD. Hence $DC \parallel EB$. Similarly, the other diagonals are parallel to their respective opposite sides. Thus, $ABPE$ is a parallelogram and $[PEB] = 1$. Letting $x = [PDC]$ and $y = [BPC] = [EDP]$, we have that $x + y = 1$. Also

$$\frac{[EDP]}{[EPB]} = \frac{DP}{PB} = \frac{[PDC]}{[PCB]} \quad \text{or} \quad \frac{y}{1} = \frac{x}{y}.$$

Eliminating x, we get $y^2 + y - 1 = 0$, so that

$$y = \frac{\sqrt{5} - 1}{2} \quad \text{and} \quad x = y^2 = \frac{3 - \sqrt{5}}{2}.$$

Then

$$\text{Area}[ABCDE] = y + x + y + 2 = (5 + \sqrt{5})/2.$$

To show that there is an infinite number of such noncongruent pentagons, construct an arbitrary triangle PDC whose area is x (as above). Now extend CP to E and DP to B so that $[EDC] = [BDC] = 1$. Finally, draw $EA \parallel BD$ and $AB \parallel EC$. It follows from the previous analysis that the constructed pentagon has the desired area property.

For another proof of the last part, consider the parallel projection of a regular pentagon with the desired area property. In rectangular coordinates, the transformation is given by $y' = ky$, $x' = x/k$, where k is an arbitrary constant. Under this transformation, areas are preserved (see NML vol. 24, pp. 10–20).

James Saxe, the top contestant of the first USAMO with a perfect score, showed more generally that the affine transformation $y' = ax + by$, $x' = cx + dy$ with determinant 1 applied to the regular pentagon constitutes all the solutions to the problem.

Comment. Möbius, in his book on the Observatory of Leipzig (p. 61), posed the more general problem of finding the area of $ABCDE$, given that $[ABC] = a$, $[BCD] = b$, $[CDE] = c$, $[DEA] = d$, and $[EAB] = e$. Gauss solved this problem and wrote his solution in the margins of the book. His solution and alternate ones are given in *Crux Mathematicorum*, 3, (1977) pp. 238–240 and 4, (1978) pp. 17–18.

It is shown that the area of $ABCDE$ is a root of the quadratic equation $t^2 - pt + q = 0$, where $p = a + b + c + d + e$ and $q = ab + bc + cd + de + ea$.

P.G./7 (1974/5). Consider the two triangles $\triangle ABC$ and $\triangle PQR$ shown in Figure 1. In $\triangle ABC$, $\angle ADB = \angle BDC = \angle CDA = 120°$. Prove that $x = u + v + w$.

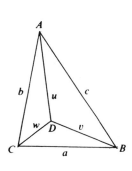
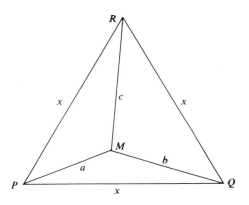

Figure 1

Solution. We shall construct the equilateral $\triangle PQR$ and show that its sides have length $u + v + w$.

Rotate $\triangle BCD$ through $60°$ counterclockwise about B to $\triangle BFE$, see Figure 2.

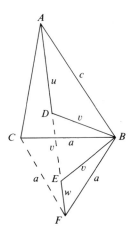
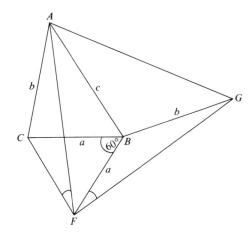

Figure 2   Figure 3

First note that $\triangle BDE$ and $\triangle BCF$ are equilateral, so that $DE = v$ and $CF = a$. Now $\angle ADE$ and $\angle DEF$ are straight angles, both $120° + 60°$, so that $AF = u + v + w$. Construct the equilateral triangle $\triangle AFG$ on AF as shown in Figure 3. Now $AF = FG$, $CF = BF$, and $\angle CFA = \angle BFG$, both $60° - \angle AFB$. Thus, $\triangle CFA \simeq \triangle BFG$. It follows that $BG = AC = b$, so that $\triangle AFG$ is the desired equilateral triangle with side length $x = u + v + w$.

Comment. There are other known interesting properties of the configurations of Figure 1 which we now note without proof. For derivations, see [95, pp. 218–222].

The point D is called an *isogonic center* of triangle ABC and can be constructed by drawing equilateral triangles BCP_1, CAP_2, and ABP_3 externally as shown in Figure 4.

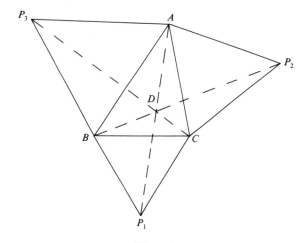

Figure 4

The segments AP_1, BP_2, and CP_3 are of equal length (this follows from the previous proof) and are concurrent at point D. It is assumed here that the largest angle of triangle ABC is less than $120°$. D also has the property that it is the point from which the sum of its distances to A, B, C is less than that from any other point. The side x is given by

$$2x^2 = a^2 + b^2 + c^2 + 4[ABC]\sqrt{3},$$

where $[ABC]$ is the area of $\triangle ABC$.

There is also another equilateral triangle which can be constructed having the point M exterior to it. Its side x' is given by

$$2x'^2 = a^2 + b^2 + c^2 - 4[ABC]\sqrt{3}.$$

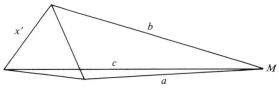

Figure 5

Finally, we leave it as an exercise to show that in Figure 3,

$$\angle ABG - \angle CAB = \angle ABF - \angle ABC = \angle FBG - \angle BCA = 60°.$$

Solid Geometry

S.G./1. (1979/2). Let S be a great circle with pole P. On any great circle through P, two points A and B are chosen equidistant from P. For any *spherical triangle ABC* (the sides are great circle arcs), where C is on S, prove that the great circle arc CP is the angle bisector of angle C.

Note. A *great circle* on a sphere is one whose center is the center of the sphere. A *pole* of the great circle S is a point P on the sphere such that the diameter through P is perpendicular to the plane of S.

Solution. The angle between two great circle arcs CA and CB drawn through a point C is measured by the size of the lune which those arcs, when extended, cut off on the sphere. It is therefore natural to look at the sphere from above the pole P as in the figure. A proof follows by symmetry.

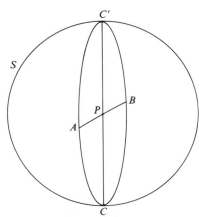

The great circle arcs *CA*, *CP*, and *CB* when extended will meet at a point *C'* which is antipodal (diametrically opposite) to *C*. Now consider a rotation of the sphere 180° about *OP* as an axis. This rotation sends every great circle through *P* on itself with only the orientation reversed, so that *B* falls somewhere on the great circle arc through *P* and *A*. Since arcs *AP* and *PB* are congruent, *B* and *A* are interchanged by the rotation, as are the points *C* and *C'*. Thus lune *ACPC'* is mapped onto lune *BC'PC*, the two lunes are congruent, and *CC'* bisects angles *C* and *C'*.

Note that arc *CA* + arc *AC'* = arc *CA* + arc *CB* = 1/2 a great circle. Hence the great circle *S* is a *spherical ellipse* with foci *A* and *B*. This result is used in the solution of problem B-4 of the Putnam Competition 1971: Determine the entire class of real spherical ellipses which are circles [20, p. 16, 73–74]. A spherical ellipse with foci *A*, *B* on a given sphere is defined as the set of all points *P* on the sphere such that arc *PA* + arc *PB* = constant. Here arc *PA* denotes the shortest distance on the sphere (a minor great circle arc) between *P* and *A*.

S.G./2. (1972/2). A given tetrahedron *ABCD* is isosceles, that is, *AB* = *CD*, *AC* = *BD*, *AD* = *BC*. Show that the faces of the tetrahedron are acute-angled triangles.

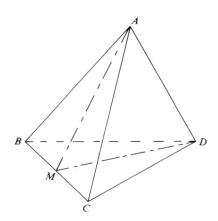

First solution. It follows from the hypothesis that the four faces are congruent and that the trihedral angle at each vertex is made up of the three distinct angles of a face.

Let *M* be the midpoint of *BC*. By the triangle inequality,

$$AM + MD > AD = BC = 2MC.$$

Triangles *ABC* and *DCB* are congruent, so *AM* = *DM*. Thus

$$2MD > 2MC;$$

that is, *MD* is greater than the radius of the circle in the plane of *BCD* with diameter *BC*. Therefore *D* lies outside this circle and angle *BDC* is acute. The same argument applies to every face angle.

Second solution. We saw above that the sum of the three face angles at each vertex of our tetrahedron is 180°. That each face angle is acute now follows from the

> **THEOREM.** *The sum of any two angles of a trihedral angle is greater than the third angle.*

For completeness, we give two proofs of this basic inequality.

Since $\cos x$ is a decreasing function for x in $[0, 180°]$, the inequality stated in the theorem is equivalent to

$$(1) \qquad \cos DAC > \cos(DAB + BAC)$$
$$= \cos DAB \cos BAC - \sin DAB \sin BAC.$$

We introduce the vectors

$$\mathbf{b} = \overrightarrow{AB}, \qquad \mathbf{c} = \overrightarrow{AC}, \qquad \mathbf{d} = \overrightarrow{AD}$$

and use the definitions of dot product and vector product of two vectors to rewrite inequality (1) in the form

$$(1)' \qquad (\mathbf{b} \cdot \mathbf{b})(\mathbf{d} \cdot \mathbf{c}) > (\mathbf{b} \cdot \mathbf{d})(\mathbf{b} \cdot \mathbf{c}) - |\mathbf{b} \times \mathbf{d}||\mathbf{b} \times \mathbf{c}|.$$

Since

$$(\mathbf{b} \cdot \mathbf{d})(\mathbf{b} \cdot \mathbf{c}) - (\mathbf{b} \cdot \mathbf{b})(\mathbf{d} \cdot \mathbf{c}) = \mathbf{b} \cdot \{\mathbf{c}(\mathbf{b} \cdot \mathbf{d}) - \mathbf{b}(\mathbf{c} \cdot \mathbf{d})\}$$
$$= \mathbf{b} \cdot \{\mathbf{d} \times (\mathbf{c} \times \mathbf{b})\}$$
$$= (\mathbf{b} \times \mathbf{d}) \cdot (\mathbf{c} \times \mathbf{b}),$$

and since

$$|\mathbf{b} \times \mathbf{d}||\mathbf{b} \times \mathbf{c}| > (\mathbf{b} \times \mathbf{d}) \cdot (\mathbf{c} \times \mathbf{b}),$$

the desired inequality (1)′ follows.

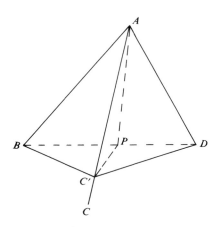

We give a second, elementary geometric proof of the theorem. In the figure on the previous page, let $\angle BAD$ be the largest face angle of the trihedral angle at vertex A. It now suffices to show that

$$\angle BAC + \angle CAD > \angle DAB.$$

We can choose P on BD such that $\angle BAP = \angle BAC$. We then choose C' on AC such that $AC' = AP$. It follows that $\triangle BAP \cong \triangle BAC'$ and so $BC' = BP$. Also, $BC' + C'D > BD = BP + PD$, so that $C'D > PD$. Now by considering triangles DAP and DAC' and applying the theorem "if two triangles have two sides of one equal respectively to two sides of the other, but the included angles are unequal, then the larger of these angles is opposite to the larger third side and conversely", we get $\angle C'AD > \angle PAD$. Finally

$$\angle BAC' + \angle C'AD = \angle BAP + \angle C'AD > \angle BAP + \angle PAD = \angle BAD.$$

Another proof using the duality with a spherical triangle is given in the solution of the subsequent problem S.G./5.

S.G./3. (1977/4). Prove that if the opposite sides of a skew (non-planar) quadrilateral are congruent, then the line joining the midpoints of the two diagonals is perpendicular to these diagonals, and conversely, if the line joining the midpoints of the two diagonals of a skew quadrilateral is perpendicular to these diagonals, then the opposite sides of the quadrilateral are congruent.

Solution. For a problem involving equal lengths and perpendicularity, vector methods furnish a very direct solution. Denoting the quadrilateral by $ABCD$, we introduce vectors $\mathbf{a}, \mathbf{b}, \mathbf{c}, \mathbf{d}$ from some common origin to the vertices A, B, C, D and note that the midpoints of the diagonals are $\frac{1}{2}(\mathbf{a} + \mathbf{c})$, $\frac{1}{2}(\mathbf{b} + \mathbf{d})$ and the vector joining them is $\frac{1}{2}[(\mathbf{a} + \mathbf{c}) - (\mathbf{b} + \mathbf{d})]$.

We can now rephrase the problem in the form: Given that

(1) $(\mathbf{a} - \mathbf{b}) \cdot (\mathbf{a} - \mathbf{b}) = (\mathbf{c} - \mathbf{d}) \cdot (\mathbf{c} - \mathbf{d})$

(2) $(\mathbf{b} - \mathbf{c}) \cdot (\mathbf{b} - \mathbf{c}) = (\mathbf{a} - \mathbf{d}) \cdot (\mathbf{a} - \mathbf{d}),$

show that

(3) $[(\mathbf{a} + \mathbf{c}) - (\mathbf{b} + \mathbf{d})] \cdot (\mathbf{a} - \mathbf{c}) = 0,$

(4) $[(\mathbf{a} + \mathbf{c}) - (\mathbf{b} + \mathbf{d})] \cdot (\mathbf{b} - \mathbf{d}) = 0,$

and conversely. Subtracting (2) from (1) gives (3), while adding (2) to (1) gives (4). Conversely, adding (3) and (4) gives (1), and subtracting (3) from (4) gives (2).

S.G./4. (1983/4). Six segments S_1, S_2, S_3, S_4, S_5 and S_6 are given in a plane. These are congruent to the edges AB, AC, AD, BC, BD and CD, respectively, of a tetrahedron $ABCD$. Show how to construct a segment congruent to the altitude of the tetrahedron from vertex A with straight-edge and compasses.

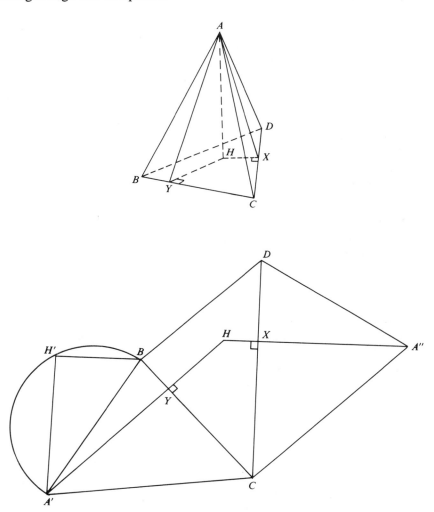

Solution. In the figure, AH denotes the desired altitude. Now consider the plane through AH perpendicular to CD. This plane is perpendicular to the plane BCD and intersects plane ACD in a line AX. It follows that CD is perpendicular to both HX and AX. Similarly, the plane through AH perpendicular to BC intersects plane ABC in a line AY, and BC is perpendicular to HY and AY. Note that H can be obtained as the intersection of two lines in plane BCD. The first is the intersection of the plane through $A \perp$ to CD with plane BCD, the second is the intersection of the plane through $A \perp$ to BC with plane BCD. We now develop the tetrahedron into a plane by cutting along the edges AB, AC, and AD. We lay out the triangles ABC and ACD into the plane of BCD as in the second figure on p. 59. The vertices A of $\triangle ABC$ and $\triangle ACD$ go into points A', A'' respectively in the plane of BCD. All these face triangles can be constructed since their sides are known. Point H is now determined as the intersection of the lines $A''X$ perpendicular to CD and $A'Y$ perpendicular to BC. Referring to the first figure, AHB is a right triangle with AB as hypotenuse and BH and AH as the other sides. Thus the desired altitude AH is obtained as $A'H'$ in the second figure by constructing a semicircle on $A'B$ and marking off $BH' = BH$.

S.G./5. (1981/4). The sum of the measures of all the face angles of a given convex polyhedral angle is equal to the sum of the measures of all its dihedral angles. Prove that the polyhedral angle is a trihedral angle.

Note. A convex polyhedral angle may be formed by drawing rays from an exterior point to all points of a convex polygon.

Solution. A dual and often useful way in dealing with problems concerning polyhedral angles P is to consider the associated spherical polygon P' formed by the intersection of P with a unit sphere centered at the vertex of P. (This is treated in books on spherical trigonometry.) If P is convex, so then is P'. The sides and angles of P' will be equal to the corresponding face angles and dihedral angles of P. In particular for a trihedral angle, we would obtain a spherical triangle. Since the sum of two sides of a spherical triangle is greater than the third side (as for plane triangles), this gives the dual result (derived differently in the solution of S.G./2) that the sum of any two angles of a trihedral angle is greater than the third angle.

It now follows that the given problem is equivalent to determining all convex spherical polygons such that the sum of the measures of all the sides is equal to the sum of the measures of all the angles. Since for this problem P' is convex, the sum of its sides cannot exceed a great circle or $360°$. Dually, this implies that the sum of the face angles of the given P cannot exceed $360°$ (at the end we will also give a direct geometric proof of this). The sum of the angles of a spherical triangle always exceeds $180°$ (the difference between the sum of the angles and 180° is known as the "spherical excess" and is proportional to the area of the triangle). If P' has n sides, it can be divided into $n - 2$ spherical triangles; hence the sum of the angles of P' exceeds $(n - 2)180°$. Thus we must have $(n - 2)180° < 360°$, and so n is at most 3. To see that 3 is possible, just consider a trirectangular triangle. Each side and each angle equals 90°.

We now give an alternate proof that the sum of the face angles of a convex polyhedral angle is less than 360°. For simplicity, we use a 4-hedral angle. However, the proof carries through for an n-hedral angle. Let P-$ABCD$ be a convex 4-hedral angle and let Q be any interior point of the plane polygon $ABCD$ as in the figure.

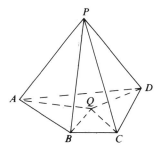

Since the sum of two angles of a trihedral angle is greater than the third angle, we obtain the following inequalities by considering the 4 base trihedral angles:

$$\angle PBA + \angle PBC > \angle ABC,$$
$$\angle PCB + \angle PCD > \angle BCD,$$
$$\angle PDC + \angle PDA > \angle CDA,$$
$$\angle PAD + \angle PAB > \angle DAB.$$

We add these four inequalities. We also add the equalities

$$\angle PBA + \angle PAB = 180° - \angle APB$$
$$\angle PBC + \angle PCB = 180° - \angle BPC$$
$$\angle PCD + \angle PDC = 180° - \angle CPD$$
$$\angle PDA + \angle PAD = 180° - \angle DPA,$$

which just say that the sum of the angles of a triangle is 180°. When we substitute the sum of the right members of the equalities for the sum of the left numbers of the inequalities above, we obtain

$$720° - (\angle APB + \angle BPC + \angle CPD + \angle DPA)$$
$$> \angle ABC + \angle BCD + \angle CDA + \angle DAB = 360°$$

so

$$360° > \angle APB + \angle BPC + \angle CPD + \angle DPA.$$

S.G./6. (1978/4).   (a) Prove that if the six dihedral angles (i.e., angles between pairs of faces) of a given tetrahedron are congruent, then the tetrahedron is regular.

(b) Is a tetrahedron necessarily regular if five dihedral angles are congruent?

First solution of (a).   We use the duality discussed in previous problem S.G./5. Consider a sphere centered at one of the vertices, say vertex A. The trihedral angle at A intersects the sphere in a spherical triangle whose angles are the dihedral angles at A, equal by hypothesis. A spherical triangle is uniquely determined by its angles. Therefore an equiangular triangle is equilateral, and so the three face angles of the trihedral angle at A are equal. Let us label each of these angles as α. Similarly, the three face angles at vertex B, vertex C, and vertex D are equal and we label them β, γ, and δ, respectively. Since the sum of all the face angles of the given tetrahedron $ABCD$ is $4 \times 180° = 720°$, we have $3(\alpha + \beta + \gamma + \delta) = 720°$ or $\alpha + \beta + \gamma + \delta = 240°$. Since the sum of each three of α, β, γ, δ is $180°$ (corresponding to the sum of the angles of each face), we have $\alpha = \beta = \gamma = \delta = 60°$. Hence all the faces of the tetrahedron are congruent equilateral triangles, and the tetrahedron is regular.

Solution of (b).   Consider, for any tetrahedron, the outward unit vectors $\mathbf{k}, \mathbf{l}, \mathbf{m}, \mathbf{n}$ normal to the four faces. Note that the angle between any two of these is the supplement of the dihedral angle formed by the corresponding faces. Thus, in a regular tetrahedron, these unit vectors end at equally spaced points on the unit sphere, that is, the six great circle arcs between pairs of points, subtended by equal angles θ are equally long. [It is easy to show that $\theta = \arccos -\frac{1}{3}$.] If we can construct four points on the unit sphere such that 5 of the 6 distances between pairs are equal but the sixth is different, then the planes tangent to the sphere at these points will form a tetrahedron with 5 and only 5 equal dihedral angles.

We accomplish this by displacing our equally spaced points slightly, so that the new points K', L', M' again form an equilateral spherical triangle, but with sides of length $\theta' < \theta$. We construct N' so that it has distance θ' from K' and L'. Then its distance from M' is *not* θ', since by (a), six equal distances would imply that the tetrahedron is regular, with normals spaced θ apart. We conclude that the answer to the question in (b) is no.

We now give a second solution of part (a).

Second solution of (a).   Let O be the center of the inscribed sphere. Let A', B', C', D' be the points of contact of this sphere with the faces opposite A, B, C, D. The angle between any two of the vectors $\mathbf{a}', \mathbf{b}', \mathbf{c}', \mathbf{d}'$ from O to A', B', C', D' is the supplement of the dihedral angle formed by the corresponding faces and hence these angles are all equal.

Since the lengths of the vectors are also equal (to the radius of the sphere), the six distances $A'B'$, $A'C'$, ..., $C'D'$ are all the same. In other words $A'B'C'D'$ is a regular tetrahedron.

To see that the tetrahedron $ABCD$ itself is regular, we observe that the faces of the latter are the tangent planes of the sphere through A', B', C', D' at these points. Hence the faces of tetrahedron $ABCD$ are invariant under all the rotations which leave the regular tetrahedron $A'B'C'D'$ unchanged, and so $ABCD$ too is regular.

S.G./7. (1980/4). The inscribed sphere of a given tetrahedron touches all four faces of the tetrahedron at their respective centroids. Prove that the tetrahedron is regular.

First solution. Let G_1, G_2 be the respective centroids and DG_1M_1, DG_2M_2 the respective medians of the triangles DAB and DBC, as in the figure.

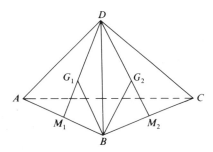

Since tangents to a sphere from an external point are equal, $DG_1 = DG_2$ and $BG_1 = BG_2$. Hence, $\triangle DBG_1 \cong DBG_2$, so that $\angle BDG_1 = \angle BDG_2$. Since $DM_1 = 3DG_1/2$ and $DM_2 = 3DG_2/2$, it follows that $\triangle DBM_1 \cong \triangle DBM_2$. Thus, $BM_1 = BM_2$, and so $BA = BC$. In a similar way, we find that any two adjacent edges are equal; hence the tetrahedron is regular.

Second solution. Let $\mathbf{a}, \mathbf{b}, \mathbf{c}, \mathbf{d}$ be vectors from the center of the inscribed sphere to the vertices A, B, C, D, respectively. The vector to the centroid G of $\triangle ABC$ is given by $\mathbf{g} = \frac{1}{3}(\mathbf{a} + \mathbf{b} + \mathbf{c})$. By hypothesis, $|\mathbf{g}| = r$, the radius of the inscribed sphere, so

$$|\mathbf{a} + \mathbf{b} + \mathbf{c}|^2 = 3\mathbf{g} \cdot (\mathbf{a} + \mathbf{b} + \mathbf{c}) = 9r^2,$$

or

(1) $$\mathbf{g} \cdot \mathbf{a} + \mathbf{g} \cdot \mathbf{b} + \mathbf{g} \cdot \mathbf{c} = 3r^2.$$

Also by hypothesis, $\mathbf{g}$ is perpendicular to the plane ABC and therefore to every line in ABC. Hence the scalar products $\mathbf{g} \cdot (\mathbf{a} - \mathbf{b})$, $\mathbf{g} \cdot (\mathbf{b} - \mathbf{c})$ vanish, yielding $\mathbf{g} \cdot \mathbf{a} = \mathbf{g} \cdot \mathbf{b} = \mathbf{g} \cdot \mathbf{c}$. From this and (1), it follows that $\mathbf{g} \cdot \mathbf{a} = \mathbf{g} \cdot \mathbf{b} = \mathbf{g} \cdot \mathbf{c} = r^2$, or

(2) $(\mathbf{a} + \mathbf{b} + \mathbf{c}) \cdot \mathbf{a} = (\mathbf{a} + \mathbf{b} + \mathbf{c}) \cdot \mathbf{b} = (\mathbf{a} + \mathbf{b} + \mathbf{c}) \cdot \mathbf{c} = 3r^2.$

Analogous results hold for the other faces; e.g. for ABD, we have

(3) $(\mathbf{a} + \mathbf{b} + \mathbf{d}) \cdot \mathbf{a} = (\mathbf{a} + \mathbf{b} + \mathbf{d}) \cdot \mathbf{b} = (\mathbf{a} + \mathbf{b} + \mathbf{d}) \cdot \mathbf{d} = 3r^2.$

From the first equalities in (2) and (3), we find that $\mathbf{a} \cdot \mathbf{c} = \mathbf{a} \cdot \mathbf{d}$; and similarly, we obtain

$$\mathbf{a} \cdot \mathbf{b} = \mathbf{a} \cdot \mathbf{c} = \mathbf{a} \cdot \mathbf{d} = \mathbf{b} \cdot \mathbf{c} = \mathbf{b} \cdot \mathbf{d} = \mathbf{c} \cdot \mathbf{d} = \lambda^2.$$

Now from (2), $|\mathbf{a}|^2 + \mathbf{a} \cdot \mathbf{b} + \mathbf{a} \cdot \mathbf{c} = 3r^2$, so $|\mathbf{a}|^2 = 3r^2 - 2\lambda^2$. Similarly, $|\mathbf{b}|^2 = |\mathbf{c}|^2 = |\mathbf{d}|^2 = 3r^2 - 2\lambda^2$. The square of the length of edge AB is

$$|\mathbf{b} - \mathbf{a}|^2 = |\mathbf{a}|^2 + |\mathbf{b}|^2 - 2\mathbf{a} \cdot \mathbf{b} = 6r^2 - 6\lambda^2.$$

Similarly, all edges have the same length and so the tetrahedron is regular.

Comment. The given problem can be extended by having the inscribed sphere touch the four faces of the tetrahedron at either their respective (i) incenters, (ii) orthocenters, or (iii) circumcenters instead of their centroids. As exercises, show that for cases (i) and (ii), the tetrahedron must still be regular. For case (iii), show that the tetrahedron must be isosceles (i.e., opposite pairs of edges are congruent).

S.G./8. (1982/5). A, B, and C are three interior points of a sphere S such that AB and AC are perpendicular to the diameter of S through A, and so that two spheres can be constructed through A, B, and C which are both tangent to S. Prove that the sum of their radii is equal to the radius of S.

First solution.   Our first solution is an analytic geometric one. As usual it is very worthwhile to pick an appropriate coordinate system in order to simplify the calculations. Any sphere S' through A, B, C must contain the circumcircle of $\triangle ABC$. Thus the center O' of S' lies on a line through the circumcenter D of $\triangle ABC$ and perpendicular to its plane. Consequently, we pick our coordinate system such that sphere S is centered at the origin O, the x-axis is along OA and O', D lie in the x, y plane. Take the radius of S be unity; then the equation of S is $x^2 + y^2 + z^2 = 1$. Now let A, D, O' have respective coordinates $(a, 0, 0)$, $(a, d, 0)$, and $(t, d, 0)$, where a, d are given constants. The equation of S' will then be $(x - t)^2 + (y - d)^2 + (z)^2 = r^2$ where t and r (radius of S') are to be determined. Since S' must contain point A,

(1) $r^2 = (a - t^2) + d^2.$

Since S' must be tangent to S, the distance between their centers must equal $1 - r$ (the difference of the radii):

(2) $$(1 - r)^2 = t^2 + d^2.$$

We now show that there are only two solutions r_1, r_2 for r and two solutions t_1, t_2 for t, where $r_1 + r_2 = 1$, and $t_1 + t_2 = a$. Subtracting (1) from (2), gives $1 - 2r = 2at - a^2$, so

(3) $$t - a = (1 - 2r - a^2)/2a.$$

Substituting from (3) into (1), we get

$$r^2(1 - a^2) - r(1 - a^2) + a^2d^2 + (1 - a^2)^2/4 = 0.$$

Thus the sum of the two roots $r_1 + r_2 = 1$. Also from (3), we get $t_1 + t_2 = a$.

Second solution. Our second solution is geometric. Consider a cross section of S and the two internally tangent spheres by a plane determined by OA and the circumcenter D of $\triangle ABC$ as in the following figure.

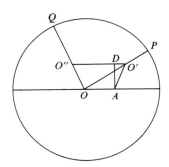

As before the centers O', O'' of the two spheres through A, B, C lie on a line through D and parallel to OA. Let P and Q be the points of tangency of the two spheres with S. The points must satisfy

$$AO' = O'P = r_1, \qquad AO'' = O''Q = r_2,$$

(4)
$$OO' = r - r_1 \quad \text{or} \quad OO' + AO' = r,$$
$$OO'' = r - r_2 \quad \text{or} \quad OO'' + AO'' = r$$

(here r, r_1 and r_2 are the three radii). By (4), O' and O'' lie on an ellipse with foci O and A. Since $O'O'' \parallel OA$, it follows by the symmetry of the ellipse that the trapezoid $OAO'O''$ is isosceles. Thus $OO' = AO''$, and finally

$$r_1 + r_2 = O'P + OO' = r.$$

Geometric Inequalities

G.I./1. (1979/4). Show how to construct a chord BPC of a given angle A through a given point P such that $1/BP + 1/PC$ is a maximum.

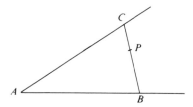

Solution. Draw PC' parallel to AB and $C'P'$ parallel to BC as in the following figure:

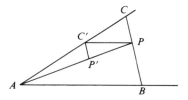

Since $AP'C'$ is similar to APC, and $PC'P'$ is similar to ABP, we have

$$\frac{P'C'}{PC} = \frac{AP'}{AP} \quad \text{and} \quad \frac{P'C'}{BP} = \frac{P'P}{AP}.$$

Adding, we get

$$\frac{P'C'}{BP} + \frac{P'C'}{PC} = \frac{(AP' + P'P)}{AP} = 1.$$

Thus

$$1/BP + 1/PC = 1/P'C'.$$

Maximizing the quantity on the left is equivalent to minimizing $P'C'$. Since $C'P'$ is a minimum when it is perpendicular to AP, BC is to be constructed perpendicular to AP.

Comment. This problem is a special case of the more general problem of finding the maximum or minimum value of $F(AB, AC, BP, PC)$, where the Angle A and the function F are given; a number of these problems have appeared in other competitions. For example:

(1) Minimize $BP \cdot PC$ (just make ABC isosceles [21; pp. 27, 95]);
(2) minimize area of ABC (just draw APA' with $PA' = PA$ and complete the parallelogram of which AA' is a diagonal and the sides are parallel to AB and AC;
(3) minimize perimeter of ABC [Crux Math. 7 (1981) 105–106];
(4) minimize BC [Crux Math. 6 (1980) 260–262].

G.I./2. (1975/4). Two given circles intersect in two points P and Q. Show how to construct a segment AB passing through P and terminating on the two circles such that $AP \cdot PB$ is a maximum.

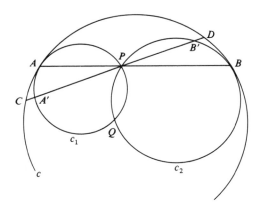

First solution. Let c_1, c_2 be the two circles. We first show that if APB is such that there is a circle c touching c_1 at A and c_2 at B, then AB is the segment giving the required maximum; and then we show how to construct A and B.

Let $A'B, PB'$ be another pair of collinear chords of c_1 and c_2. Let C, D be points where the extensions of these chords intersect the circle c. Then $CP \cdot PD = AP \cdot PB$, and hence $A'P \cdot PB' < AP \cdot PB$, as claimed.

In the figure on p. 68, we see that a circle c tangent to c_1 at A and to c_2 at B exists if $\alpha = \beta$, where α and β are half the central angles subtending chords AP and PB.

Our construction is as follows: Let O be the fourth vertex of the parallelogram with vertices O_1, P and O_2. Let A be the intersection of OO_1 with c_1, and let B be the intersection of OO_2 with c_2. We see that a circle c with center O whose radius is the sum of the radii of c_1 and c_2 touches these circles at A and B respectively. That A, P and B are collinear follows from the similarity of triangles AO_1P and AOB.

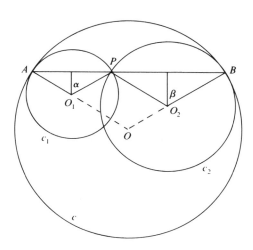

Second solution. Since $AP = 2r_1 \sin\alpha$ and $BP = 2r_2 \sin\beta$, where r_1, r_2 are the radii of c_1, c_2, respectively, we want to maximize $\sin\alpha\sin\beta$. We note that since $\angle O_1PO_2$ is fixed, so also is the sum $\alpha + \beta$. Now

$$2\sin\alpha\sin\beta = \cos(\alpha - \beta) - \cos(\alpha + \beta),$$

and since cosine is a decreasing function, the maximum occurs when $\alpha = \beta$. This implies that $AO_1 \parallel PO_2$ and $BO_2 \parallel PO_1$, and the rest follows as before.

G.I./3. (1981/3). If A, B, C are the angles of a triangle, prove that

$$-2 \leqslant \sin 3A + \sin 3B + \sin 3C \leqslant 3\sqrt{3}/2,$$

and determine when equality holds.

Solution. We can assume throughout and without loss of generality that $A \leqslant B \leqslant C$.

(i) For the lower bound, first note that since $A \leqslant 60°$, $\sin 3A \geqslant 0$. Also, $\sin 3B \geqslant -1$ and $\sin 3C \geqslant -1$. Thus

$$\sin 3A + \sin 3B + \sin 3C \geqslant -2.$$

For equality, we must have $\sin 3A = 0$, $\sin 3B = \sin 3C = -1$ so that the only case of equality is for the degenerate isosceles triangle for which $A = 0$, and $B = C = 90°$.

(ii) For the upper bound, note that $3\sqrt{3}/2 > 2$. Each of the terms $\sin 3A, \sin 3B, \sin 3C$ is at most 1; all three terms must be positive in order that their sum be a maximum. Since $A + B + C = 180°$, we must have $0° < A \leqslant B < 60°$ and $120° < C < 180°$. Now let $D = C - 120°$ so that $3A + 3B + 3D = 180°$. By Jensens's inequality for convex functions, see Glossary, it follows that

$$(1) \quad \sin 3A + \sin 3B + \sin 3D \leqslant 3 \sin \frac{3A + 3B + 3D}{3} = 3 \sin 60° = \frac{3\sqrt{3}}{2},$$

with equality *if and only if* $3A = 3B = 3D = 60°$ or $A = B = 20°$, and $C = 140°$. Note the geometric implication of (1): Of all triangles which can be inscribed in a given circle, the equilateral one has the maximum perimeter. (Here the angles $3A, 3B, 3D$ are replaced by A, B, C.)

We now give a generalization† of the upper bound inequality to

$$(2) \quad |qr \sin nA + rp \sin nB + pq \sin nC| \leqslant \left(p^2 + q^2 + r^2 \right) \frac{\sqrt{3}}{2},$$

where $p, q, r \geqslant 0$, and n is a positive integer. There is equality *iff* $p = q = r$ and $\sin nA = \sin nB = \sin nC = \pm \sqrt{3}/2$. Our starting point is the inequality‡

$$(3) \quad x^2 + y^2 + z^2 \geqslant (-1)^{m+1}(2yz \cos mA + 2zx \cos mB + 2xy \cos mC),$$

with equality *iff* $x/\sin nA = y/\sin nB = z/\sin nC$. Here x, y, z, are arbitrary real numbers and A, B, C are angles of a triangle. (3) is an easy consequence of the obvious inequality

$$\{x + (-1)^m(y \cos mC + z \cos mB)\}^2 + (y \sin mC - z \sin mB)^2 \geqslant 0.$$

In particular for m even, say $m = 2n$, we have $\cos mA = 1 - 2 \sin^2 nA$, etc., and (3) becomes

$$(4) \quad (x + y + z)^2 \geqslant 4(yz \sin^2 nA + zx \sin^2 nB + xy \sin^2 nC).$$

We further particularize by assuming that x, y, z are nonnegative and setting $\sqrt{x} = p$, $\sqrt{y} = q$, and $\sqrt{z} = r$. Then we have, with cyclic sums, that

$$\frac{p^2 + q^2 + r^2}{12} \geqslant \sum \frac{q^2 r^2 \sin^2 nA}{3} \geqslant \left(\sum \frac{qr \sin nA}{3} \right)^2,$$

where the left hand inequality comes from (4), and the right hand inequality comes from applying the power mean inequality, see Glossary.

G.I./4. (1975/2). Let A, B, C, D denote four points in space and AB the distance between A and B, and so on. Show that

$$AC^2 + BD^2 + AD^2 + BC^2 \geqslant AB^2 + CD^2.$$

---

†M. S. Klamkin, Problem 715, Crux Mathematicorum 9 (1983) pp. 60–62.
‡M. S. Klamkin, Asymmetric Triangle Inequalities, Publ. Elektrotehn. Fak. Ser. Mat. Fiz., Univ. Beograd, No. 357-380 (1971) pp. 33–44.

Solution. Let $\mathbf{a}, \mathbf{b}, \mathbf{c}, \mathbf{d}$ denote vectors from an origin 0 to points A, B, C, D, respectively. Then the inequality is equivalent to

$$(\mathbf{a} - \mathbf{c})^2 + (\mathbf{b} - \mathbf{d})^2 + (\mathbf{a} - \mathbf{d})^2 + (\mathbf{b} - \mathbf{c})^2 \geqslant (\mathbf{a} - \mathbf{b})^2 + (\mathbf{c} - \mathbf{d})^2$$

or

$$(\mathbf{a} + \mathbf{b} - \mathbf{c} - \mathbf{d})^2 \geqslant 0.$$

There is equality *if and only if* $\mathbf{a} + \mathbf{b} = \mathbf{c} + \mathbf{d}$, i.e., the four points are vertices of a parallelogram.

Alternatively, let the four points have rectangular coordinates (x_i, y_i, z_i), $i = 1, 2, 3, 4$. Then we have to show that

(1)
$$(x_1 - x_3)^2 + (x_2 - x_4)^2 + (x_1 - x_4)^2 + (x_2 - x_3)^2$$
$$\geqslant (x_1 - x_2)^2 + (x_3 - x_4)^2,$$

and similar inequalities in y_i and z_i. As above, (1) is the same as

$$(x_1 + x_2 - x_3 - x_4)^2 \geqslant 0.$$

As an extension of the above result, we will show that, if the edges of a tetrahedron $ABCD$ are given by $AB = a$, $AC = b$, $AD = c$, $CD = a_1$, $BD = b_1$, $BC = c_1$, then $a + a_1$, $b + b_1$, and $c + c_1$ are sides of an *acute* triangle. That they are sides of a triangle follows easily from the following triangle inequalities for appropriate faces of the tetrahedron:

$$a_1 + b_1 > c_1, \quad a + b > c_1, \quad a_1 + b > c, \quad a + b_1 > c.$$

Adding, we get

$$2(a + a_1) + 2(b + b_1) > 2(c + c_1), \text{ etc.}$$

From our original inequality, we have that $a^2 + a_1^2$, $b^2 + b_1^2$, and $c^2 + c_1^2$ satisfy the triangle inequality. Also, by Ptolemy's inequality in space, see Glossary, $2aa_1$, $2bb_1$, and $2cc_1$ satisfy the triangle inequality. Hence, by addition,

$$(a + a_1)^2, \quad (b + b_1)^2, \quad \text{and} \quad (c + c_1)^2$$

satisfy the triangle inequality. Thus the triangle with sides $(a + a_1)$, $(b + b_1)$ and $(c + c_1)$ is acute.

G.I./5. (1976/4). If the sum of the lengths of the six edges of a trirectangular tetrahedron $PABC$ (i.e., $\angle APB = \angle BPC = \angle CPA = 90°$) is S, determine its maximum volume.

Solution. Let $PA = a$, $PB = b$, and $PC = c$ as in the figure. Then $BC^2 = b^2 + c^2$, etc., and

(1) $S = a + b + c + \sqrt{b^2 + c^2} + \sqrt{c^2 + a^2} + \sqrt{a^2 + b^2}.$

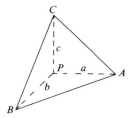

We now have to maximize the volume $V = abc/6$ subject to the constraint condition (1). By the A.M.-G.M. inequality,

$$a + b + c \geqslant 3(abc)^{1/3} \quad \text{and} \quad b^2 + c^2 \geqslant 2bc,$$

etc. Hence,

$$S \geqslant 3(abc)^{1/3} + (\sqrt{2bc} + \sqrt{2ca} + \sqrt{2ab}).$$

Applying the A.M.-G.M. inequality again to the sum in parentheses, we obtain

$$S \geqslant 3(1 + \sqrt{2})(abc)^{1/3} = 3(1 + \sqrt{2})(6V)^{1/3}.$$

Finally,

$$V_{max} = \frac{S^3}{6 \cdot 3^3(1 + \sqrt{2})^3} = \frac{(5\sqrt{2} - 7)S^3}{162},$$

and is taken on *if and only if* $a = b = c$.

The result is easily generalized by replacing the trirectangular tetrahedron by a pyramid in which the three 90° angles are changed to three angles of 2θ. Here, $V = abcn/6$† where $n^2 = 1 - 3\cos^2 2\theta + 2\cos^3 2\theta$ and

$$BC^2 = b^2 + c^2 - 2bc\cos 2\theta \geqslant 2bc(1 - \cos 2\theta) = 4bc\sin^2\theta, \quad \text{etc.}$$

---

†For a derivation of the volume of a tetrahedron *P-ABC* in terms of three concurrent edges and the angles between the edges, let

$$\alpha = \angle BPC, \beta = \angle CPA, \gamma = \angle APB, \quad \mathbf{a} = PA, \quad \mathbf{b} = PB, \quad \text{and} \quad \mathbf{c} = PC,$$

where the vector $\mathbf{a}$ has components a_1, a_2, a_3, etc. Then

$$6 \text{ Vol. } [P\text{-}ABC] = \mathbf{a} \cdot \mathbf{b} \times \mathbf{c} = \begin{vmatrix} a_1 & a_2 & a_3 \\ b_1 & b_2 & b_3 \\ c_1 & c_2 & c_3 \end{vmatrix}$$

Thus

$$36 \text{ Vol.}^2 [P\text{-}ABC] = \begin{vmatrix} a_1 & a_2 & a_3 \\ b_1 & b_2 & b_3 \\ c_1 & c_2 & c_3 \end{vmatrix} \cdot \begin{vmatrix} a_1 & b_1 & c_1 \\ a_2 & b_2 & c_2 \\ a_3 & b_3 & c_3 \end{vmatrix} = \begin{vmatrix} \mathbf{a} \cdot \mathbf{a} & \mathbf{a} \cdot \mathbf{b} & \mathbf{a} \cdot \mathbf{c} \\ \mathbf{b} \cdot \mathbf{a} & \mathbf{b} \cdot \mathbf{b} & \mathbf{b} \cdot \mathbf{c} \\ \mathbf{c} \cdot \mathbf{a} & \mathbf{c} \cdot \mathbf{b} & \mathbf{c} \cdot \mathbf{c} \end{vmatrix}$$

$$= (abc)^2 \begin{vmatrix} 1 & \cos\gamma & \cos\beta \\ \cos\gamma & 1 & \cos\alpha \\ \cos\beta & \cos\alpha & 1 \end{vmatrix} = (abc)^2 (1 - \cos^2\alpha - \cos^2\beta - \cos^2\gamma + 2\cos\alpha\cos\beta\cos\gamma).$$

Then continuing as before,

$$V_{\max} = nS^3/162(1 + 2\sin\theta)^3.$$

The given problem corresponds to the special case $2\theta = 90°$.

G.I./6. (1974/3). Two boundary points of a ball of radius 1 are joined by a curve contained in the ball and having length less than 2. Prove that the curve is contained entirely within some hemisphere of the given ball.

Solution. Let A and B denote the endpoints of the curve. Now consider the plane π that passes through the center O of the sphere and which is normal to the angle bisector of $\angle AOB$. We will show that curve AB lies in the open hemisphere formed by this plane that contains A and B. Assume to the contrary that a point X of AB lies in or below π, and let X' be the intersection point of AX and π. Then by the triangle inequality, $AX + XB \geqslant AX' + X'B$. Now consider A', the reflection of A across π (see figure). $A'B = 2$ since A', O, and B are collinear. Then $AX' + X'B = A'X' + X'B \geqslant A'B = 2$ which gives a contradiction.

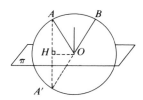

Similarly, we can show more generally that if two boundary points of a centro-symmetric body, whose minimum central diameter has length 2, are joined by a curve of length < 2, then the curve must lie in some open half of the body bounded by a plane section through the center.

For a related problem† show that any closed curve of length $< 2\pi$ on a unit sphere lies in some open hemisphere.

G.I./7. (1973/1). Two points P and Q lie in the interior of a regular tetrahedron $ABCD$. Prove that angle $PAQ < 60°$.

First solution. We can assume without loss of generality that each edge of $ABCD = 1$, that P and Q lie in the interior of $\triangle BCD$, and that line PQ intersects BC in R and CD in S as in the figure. Then $\angle PAQ < \angle RAS$.

---

†See G. D. Chakerian, M. S. Klamkin, Minimal covers for closed curves, Math. Mag. 46 (1973) pp. 55–61 and the references therein.

We now show that RS is the shortest side of $\triangle ARS$ and this implies that $\angle RAS$ (and then $\angle PAQ$) is less than $60°$. In $\triangle RSD$, $\angle RSD > 60°$ and $\angle RDS < 60°$. Thus $RD > RS$. Since $AR = RD$ (from congruent triangles BDR and BAR), $AR > RS$. Similarly, $AS > RS$. Hence RS is the shortest side of $\triangle ARS$.

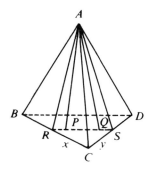
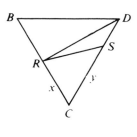

Second solution (by the law of cosines). In $\triangle RCS$, see the second figure,

$$RS^2 = x^2 + y^2 - 2xy\cos 60° = x^2 + y^2 - xy.$$

In $\triangle ACR$,

$$AR^2 = x^2 - x + 1;$$

and in $\triangle ACS$

$$AS^2 = y^2 - y + 1,$$

where $0 < x < 1$, $0 < y < 1$. Then $AR > RS$, since

$$AR^2 - RS^2 = (1 - y)(1 + y - x) > 0.$$

Similarly, $AS > RS$. Hence, $\angle PAQ < 60°$.

As a generalization, we will show that if $ABCD$ is an arbitrary tetrahedron with non-obtuse face angles at vertex A, and if P, Q are two points within or on $ABCD$, then

$$\angle PAQ \leqslant \max\{\angle BAC, \angle CAD, \angle DAB\}.$$

Our proof is vectorial. Let $\mathbf{v}$ denote a vector from origin A to a point V, etc. Without loss of generality we may choose points P, Q to lie within or on triangle BCD and take $|\mathbf{b}| = |\mathbf{c}| = |\mathbf{d}| = 1$. Then P, Q have the vector representation (barycentric coordinates)

$$\mathbf{p} = x\mathbf{b} + y\mathbf{c} + z\mathbf{d}, \qquad \mathbf{q} = u\mathbf{b} + v\mathbf{c} + w\mathbf{d},$$

where $x + y + z = 1$, $x, y, z, \geqslant 0$, $u + v + w = 1$, $u, v, w \geqslant 0$, and $|\mathbf{p}| \leqslant 1$, $|\mathbf{q}| \leqslant 1$.

Since $\cos\theta$ is a decreasing function in $[0, \pi]$, our inequality is equivalent to

$$\frac{\mathbf{p}\cdot\mathbf{q}}{|\mathbf{p}|\,|\mathbf{q}|} \geqslant \mathbf{p}\cdot\mathbf{q} = (x\mathbf{b} + y\mathbf{c} + z\mathbf{d})\cdot(u\mathbf{b} + v\mathbf{c} + w\mathbf{d})$$

$$\geqslant \min\{\mathbf{b}\cdot\mathbf{c}, \quad \mathbf{c}\cdot\mathbf{d}, \quad \mathbf{d}\cdot\mathbf{b}\}.$$

Multiplying out, we get

$$\mathbf{p}\cdot\mathbf{q} = xu + yv + zw + (yu + xv)\mathbf{b}\cdot\mathbf{c}$$
$$+ (zv + yw)\mathbf{c}\cdot\mathbf{d} + (xw + zu)\mathbf{d}\cdot\mathbf{b}.$$

Since the face angles at A are non-obtuse, $\mathbf{b}\cdot\mathbf{c}$, $\mathbf{c}\cdot\mathbf{d}$ and $\mathbf{d}\cdot\mathbf{b} \geqslant 0$, so

$$\mathbf{p}\cdot\mathbf{q} \geqslant \{xu + yv + zw + (yu + xv) + (zv + yw) + (xw + zu)\}$$
$$\times \min\{\mathbf{b}\cdot\mathbf{c}, \quad \mathbf{c}\cdot\mathbf{d}, \quad \mathbf{d}\cdot\mathbf{b}\}$$
$$= (x + y + z)(u + v + w)\min\{\mathbf{b}\cdot\mathbf{c}, \quad \mathbf{c}\cdot\mathbf{d}, \quad \mathbf{d}\cdot\mathbf{b}\}$$
$$= \min\{\mathbf{b}\cdot\mathbf{c}, \quad \mathbf{c}\cdot\mathbf{d}, \quad \mathbf{d}\cdot\mathbf{b}\}.$$

In terms of spherical triangles, the above inequality is equivalent to the result that the greatest chord of an acute spherical triangle† is the greatest side. Note that if ABC is a spherical triangle with $\angle B = \angle C = \pi/2$, then all chords of the triangle through A have the same length as AB and AC. Also, if $AB = AC > \pi/2 > BC$, then the angle bisector of $\angle A$ is longer than AB. See Figure 1. Here, $AB' = AC' = \pi/2$, and $\angle ABD > \angle ADB = \pi/2$.

Figure 1

We can extend the previous result by requiring that only two of the face angles at vertex A of the tetrahedron $ABCD$ be non-obtuse. Here we prefer to deal with the corresponding spherical triangle rather than the tetrahedron (for other examples of this duality, see the solutions to S.G./2, 5 and 6.

---

†We are assuming as usual that the triangle is proper or convex (i.e., its perimeter is less than that of a great circle).

We place vertex A of the tetrahedron at the center of a unit sphere and consider the spherical triangle formed by the intersection of the sphere with the three faces of the tetrahedron that meet at the center. To conform to the convention of labelling a triangle (spherical or plane) by letters A, B, C, we shall call the vertices of this spherical triangle A, B, C and the opposite sides (arcs) a, b, c, respectively, ignoring that previously A denoted the vertex of the tetrahedron.

Here we assume in spherical triangle ABC that the sides a, b, c satisfy $b, c \leqslant \pi/2 < a$, and we will show that any chord of the triangle is smaller than the largest side a. We use the Law of Cosines for spherical triangles (see [140; pp. 261–282] for a summary of the basic formulae and applications of spherical trigonometry), i.e.,

(1) $$\cos a = \cos b \cos c + \sin b \sin c \cos A, \text{ etc.}$$

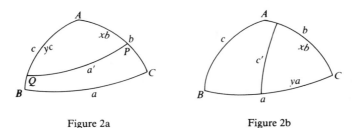

Figure 2a   Figure 2b

We may assume that the endpoints P, Q of our chord are on the perimeter of $\triangle ABC$, and that both are not on the same side. Consider first the case when P is on AC and Q is on AB. Let the arc lengths $AP = xb$, $AQ = yc$, where $0 \leqslant x \leqslant 1$, $0 \leqslant y \leqslant 1$, and let $PQ = a'$, see Figure 2a. Then

$$\cos a' = \cos xb \cos yc + \sin xb \sin yc \cos A.$$

Since

$\cos a' - \cos a$

$\quad = (\cos xb \cos yc - \cos b \cos c) - \cos A(\sin b \sin c - \sin xb \sin yc)$

$\quad \geqslant 0,$

we have $a' \leqslant a$. ($\cos A \leqslant 0$ by hypothesis and law of cosines.)

Next we treat the case when one end of the chord is on side a. Assume without loss of generality that the other end is on side b, see Figure 2b.

Here

$$\cos c' = \cos xb \cos ya + \sin xb \sin ya \cos C,$$

where $0 \leqslant x \leqslant 1, 0 \leqslant y \leqslant 1$. Since

$$\cos c' - \cos a = (\cos xb \cos ya - \cos a) + \sin xb \sin ya \cos C \geqslant 0,$$

we have $c' < a$. ($\cos C \geqslant 0$ by hypothesis and law of cosines.)

Exercise: Prove that the greatest chord of a triangle or a tetrahedron or, more generally, a simplex is the greatest edge. (Hint: This can be done by a convexity argument or by repeated adroit use of the triangle inequality and barycentric coordinates, as before; see also Amer. Math. Monthly, 68 (1961), p. 245.)

G.I./8. (1985/3). Let A, B, C and D denote any four points in space such that at most one of the distances AB, AC, AD, BC, BD and CD is greater than 1. Determine the maximum value of the sum of the six distances.

Solution. Let AD denote the distance which can be greater than 1. It is easy to see that if the other five distances are fixed, AD is a maximum when A and D are opposite vertices of a planar quadrilateral $ABDC$. For fixed positions of B and C, both A and D must lie within the intersection of the unit circles with B and C as centers. This region is centrally symmetric, so that its longest chord must pass through the midpoint O of BC. It follows that the longest chord of this region coincides with the common chord of the two unit circles; see figure.

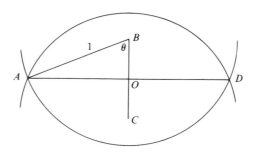

If A and D are the endpoints of this common chord, the distances AD, AB, AC, DB and DC are simultaneously maximized, the last four being equal to 1. The problem now reduces to maximizing $AD + BC$. Let

$\theta = \angle ABO$. We have $60° \leqslant \theta \leqslant 90°$, since $0 \leqslant BC \leqslant 1$. Then we can write

$$AD + BC = 2(\sin \theta + \cos \theta) = 2\sqrt{2} \sin(\theta + 45°).$$

Since this is decreasing on $60° \leqslant \theta \leqslant 90°$, the maximum is attained at the endpoint $\theta = 60°$. It follows that the maximum of

$$AD + BC + AB + AC + DB + DC$$

is

$$2(\sin 60° + \cos 60°) + 4 = 5 + \sqrt{3}.$$

G.I./9. (1984/3). P, A, B, C and D are five distinct points in space such that $\angle APB = \angle BPC = \angle CPD = \angle DPA = \theta$, where θ is a given acute angle. Determine the greatest and least values of $\angle APC + \angle BPD$.

First solution.   Let $\alpha = \angle APC$ and $\beta = \angle BPD$. Clearly, the least value of $\alpha + \beta$ is 0, attained when P, A and C are collinear and P, B and D are collinear.

In order for $\alpha + \beta$ to attain its greatest value, P-$ABCD$ must form a convex polyhedral angle. This has a dual representation as a spherical quadrilateral. Consider the intersections of a unit sphere centered at P with the planes APB, BPC, etc. We obtain a spherical rhombus $ABCD$. As in plane geometry, the diagonals AC and BD are perpendicular and bisect each other at their intersection Q, see figure. [This follows by congruent spherical triangles, e.g., $\triangle BAD \cong \triangle BCD$, and then $\triangle ABQ \cong \triangle CBQ$, etc.] Here, $\widehat{AB} = \widehat{BC} = \widehat{CD} = \widehat{DA} = \theta$, $\widehat{AC} = \alpha$ and $\widehat{BD} = \beta$.

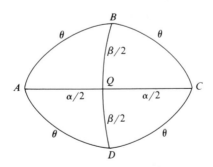

Applying the spherical law of cosines to $\triangle AQB$, we have

$$\cos \theta = \cos \frac{\alpha}{2} \cos \frac{\beta}{2} + \sin \frac{\alpha}{2} \sin \frac{\beta}{2} \cos \angle AQB = \cos \frac{\alpha}{2} \cos \frac{\beta}{2}.$$

This equation can be rewritten as

$$\cos\frac{\alpha + \beta}{2} = 2\cos\theta - \cos\frac{\alpha - \beta}{2}.$$

Since $\cos x$ is monotonically decreasing in $[0, \pi]$, $\alpha + \beta$ is a maximum when $\alpha = \beta$, with value $2\arccos(2\cos\theta - 1)$. This corresponds to the case where P is the vertex and PA, PB, PC and PD are edges of a right square pyramid.

Second solution. Let $\mathbf{a}$, $\mathbf{b}$, $\mathbf{c}$ and $\mathbf{d}$ be unit vectors along PA, PB, PC and PD, respectively. Then

(1) $$\mathbf{a} \cdot \mathbf{b} = \mathbf{b} \cdot \mathbf{c} = \mathbf{c} \cdot \mathbf{d} = \mathbf{d} \cdot \mathbf{a} = \cos\theta,$$

which implies

(2) $$(\mathbf{a} - \mathbf{c}) \cdot (\mathbf{b} - \mathbf{d}) = 0.$$

Equation (2) holds if $\mathbf{a} - \mathbf{c}$ and $\mathbf{b} - \mathbf{d}$ are perpendicular, or if at least one of these vectors is 0. If $\mathbf{a} = \mathbf{c}$, P, A and C are collinear, and $\alpha = 0$; if $\mathbf{b} = \mathbf{d}$, P, B and D are collinear, and $\beta = 0$. If α and β both vanish, we get the minimum value $\alpha + \beta = 0$, see First solution. For the maximum value of $\alpha + \beta$ all four vectors are distinct and $\mathbf{a} - \mathbf{c}$ is perpendicular to $\mathbf{b} - \mathbf{d}$.

From (1) it also follows that the unit vectors $\mathbf{a}, \mathbf{b}, \mathbf{c}, \mathbf{d}$ satisfy

$$(\mathbf{a} + \mathbf{c}) \cdot (\mathbf{a} - \mathbf{c}) = (\mathbf{a} + \mathbf{c}) \cdot (\mathbf{b} - \mathbf{d}) = 0,$$
$$(\mathbf{b} + \mathbf{d}) \cdot (\mathbf{b} - \mathbf{d}) = (\mathbf{b} + \mathbf{d}) \cdot (\mathbf{a} - \mathbf{c}) = 0.$$

This means that $\mathbf{a} + \mathbf{c}$ and $\mathbf{b} + \mathbf{d}$ are perpendicular to both, $\mathbf{a} - \mathbf{c}$ and $\mathbf{b} - \mathbf{d}$; hence $\mathbf{a} + \mathbf{c}$ and $\mathbf{b} + \mathbf{d}$ are collinear, and

$$(\mathbf{a} + \mathbf{c}) \cdot (\mathbf{b} + \mathbf{d}) = |\mathbf{a} + \mathbf{c}| |\mathbf{b} + \mathbf{d}|.$$

Multiplying out and using (1), we find that

$$4\cos\theta = 2\sqrt{(1 + \cos\alpha)(1 + \cos\beta)} = 4\cos\frac{\alpha}{2}\cos\frac{\beta}{2},$$

as in the previous solution, and we complete the argument in the same way.

G.I./10. (1982/3). If a point A_1 is in the interior of an equilateral triangle ABC and point A_2 is in the interior of $\Delta A_1 BC$, prove that

$$\text{I.Q.} (A_1 BC) > \text{I.Q.} (A_2 BC),$$

where the *isoperimetric quotient* of a figure F is defined by

$$\text{I.Q.}(F) = \text{Area}(F)/[\text{Perimeter}(F)]^2.$$

Solution.   It suffices to show that the I.Q. of a triangle DEF is an increasing function of each of the base angles E, F separately, for $0 < E < 60°, 0 < F < 60°$.

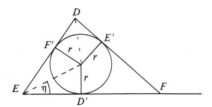

Looking at the figure, we see that the area of $\triangle DEF = \frac{1}{2}r$ (Perimeter). Hence

$$\frac{1}{2\,(\text{I.Q.})} = \frac{\text{Perimeter}}{r}$$

$$= \frac{DF'}{r} + \frac{F'E}{r} + \frac{ED'}{r} + \frac{D'F}{r} + \frac{FE'}{r} + \frac{E'D}{r}$$

$$= 2(\cot \eta + \cot \varphi + \cot \delta),$$

where η, φ, δ are the half angles of $\triangle EFD$, so $\eta + \varphi + \delta = 90°$. Since $\cot \delta = \tan(90° - \delta) = \tan(\eta + \varphi)$, we have that

$$\frac{1}{4\,(\text{I.Q.})} = \cot \eta + \cot \varphi + \tan(\eta + \varphi)$$

$$= \cot \eta + \cot \varphi + \frac{\cot \eta + \cot \varphi}{\cot \eta \cot \varphi - 1}.$$

Setting $\cot \eta = p, \cot \varphi = q$, we find that $1/4$ (I.Q.) is the function

$$J(p, q) = p + q + \frac{p + q}{pq - 1},$$

symmetric in p and q. To show that I.Q. is an increasing function of $\angle E$ $(0 < E < 60°)$ for fixed F $(0 < F < 60°)$ is equivalent to showing that $1/4$ (I.Q.) is a decreasing function of η, $0 < \eta < 30°$, for fixed φ, $0 < \varphi < 30°$, or that J is an increasing function of $p, p > \sqrt{3}$, for fixed $q > \sqrt{3}$.

A simple way of completing the proof, and one that makes no use of calculus, is to write our function in the form

(1)
$$f(u) = u + \frac{c}{u},$$

c a positive constant and use the following litte lemma (which we prove below): *A function of form (1) is increasing in the interval* $u \geq \sqrt{c}$.

To see this, take two numbers u and v in this interval with $v > u$. Then

$$f(v) - f(u) = v + \frac{c}{v} - u - \frac{c}{u} = (v - u)\frac{uv - c}{uv},$$

and this is clearly positive for $u, v \geq \sqrt{c}$.

Next we manipulate J into form (1):

$$p + q + \frac{p + q}{pq - 1} = p + q + \frac{1}{q} + \frac{1 + 1/q^2}{p - 1/q}$$

$$= p - \frac{1}{q} + \frac{1 + 1/q^2}{p - 1/q} + q + \frac{2}{q};$$

$p - 1/q$ plays the role of u, $1 + 1/q^2$ that of c, and when $q \geq \sqrt{3}$ is fixed, $q + 2/q$ is just an added "constant" which does not affect the monotonicity of J. Thus J increases for

$$p - \frac{1}{q} \geq \sqrt{1 + \frac{1}{q^2}} \quad \text{or} \quad p \geq \frac{1}{q} + \sqrt{1 + \frac{1}{q^2}},$$

and when $q \geq \sqrt{3}$, the right side is $< \sqrt{3}$, so the domain of monotonicity includes $p \geq \sqrt{3}$, as desired.

One of the more routine ways of proving that

$$J = \frac{1}{4 \,(\text{I.Q.})} = \cot \eta + \cot \varphi + \tan(\eta + \varphi)$$

decreases for $0 < \eta < \pi/6$ and fixed φ, $0 < \varphi < \pi/6$, is to show that its derivative with respect to η is negative:

$$J_\eta = -\frac{1}{\sin^2 \eta} + \frac{1}{\cos^2(\eta + \varphi)} = -\frac{1}{\cos^2\left(\dfrac{\pi}{2} - \eta\right)} + \frac{1}{\cos^2(\eta + \varphi)}.$$

For $0 < \eta < \dfrac{\pi}{6}$, $0 < \varphi < \dfrac{\pi}{6}$, we have $2\eta + \varphi < \dfrac{\pi}{2}$, so $\eta + \varphi < \dfrac{\pi}{2} - \eta$, so that

$$\frac{1}{\cos^2(\eta + \varphi)} < \frac{1}{\cos^2\left(\dfrac{\pi}{2} - \eta\right)}, \quad \text{and } J_\eta \text{ is indeed negative.}$$

In Problem 83-5 (SIAM Review (1984) pp. 275–276), I had given an extension to tetrahedra. Subsequently (SIAM Review (1986) pp. 88–89), Noam Elkies, a former

U.S.A. Olympiad winner proved the generalization to n-dimensional simplexes. By means of some n-dimensional geometry, the problem reduces to showing that

(1)
$$\frac{\cot \theta_1 + \cot \theta_2 + \cdots + \cot \theta_n}{\left(\cot \tfrac{1}{2}\theta_1 + \cot \tfrac{1}{2}\theta_2 + \cdots + \cot \tfrac{1}{2}\theta_n \right)^n}$$

is an increasing function of each of the angles θ_i for $0 < \theta_i < \arccos 1/n$. For $n = 2$, θ_1 and θ_2 correspond to the angles E and F above. For $n = 3$, $\theta_1, \theta_2,$ and θ_3 correspond to the three dihedral angles at the base of the tetrahedron.

Note that there is a corresponding result for simplexes S' of vertices $A', B_1, B_2, \ldots, B_n$ which contain a regular simplex S of vertices $A, B_1, B_2, \ldots, B_n$. It then follows that (1) is a decreasing function of the "base angles" for $\theta_i > \arccos 1/n$.

Inequalities

I/1. (1974/2).   Prove that if a, b, and c are positive real numbers, then

$$a^a b^b c^c \geqslant (abc)^{(a+b+c)/3}.$$

Solution.   First we show that

(1)
$$x^x y^y \geqslant x^y y^x$$

for $x > 0$, $y > 0$. We can assume without loss of generality that $x \geqslant y$ (and also $a \geqslant b \geqslant c$). Since (1) is equivalent to $(x/y)^{x-y} \geqslant 1$ it is obviously true. Then by multiplying the three inequalities

$$a^a b^b \geqslant a^b b^a, \qquad b^b c^c \geqslant b^c c^b, \qquad c^c a^a \geqslant c^a a^c,$$

we get

$$\left(a^a b^b c^c \right)^2 \geqslant a^{b+c} b^{c+a} c^{a+b},$$

and then that $(a^a b^b c^c)^3 \geqslant (abc)^{a+b+c}$, which is equivalent to the desired result. Equality holds if and only if $a = b = c$.

More generally, it follows from (1) in a similar way that

(2)
$$\left(a_1^{a_1} a_2^{a_2} \cdots a_n^{a_n} \right)^n \geqslant (a_1 a_2 \cdots a_n)^{a_1 + a_2 + \cdots + a_n}$$

where the a_i's are positive real numbers.

Also, (2) can be obtained by applying Jensen's inequality [121; p. 23] for the convex function $\ln x^x$ (for $x > 0$) and then the A.M.–G.M. inequality applied to the argument of this monotone function, i.e.,

$$\frac{1}{n} \sum \ln a_i^{a_i} \geqslant \ln \left[\frac{1}{n} \sum a_i \right]^{(\Sigma a_i)/n} \geqslant \ln \left(\prod a_i \right)^{(\Sigma a_i)/n^2}.$$

I/2. (1978/1).   Given that a, b, c, d, e are real numbers such that

$$a + b + c + d + e = 8,$$
$$a^2 + b^2 + c^2 + d^2 + e^2 = 16.$$

Determine the maximum value of e.

Solution.   It follows from the power mean inequality† or Cauchy's inequality† that

$$4(a^2 + b^2 + c^2 + d^2) \geq (a + b + c + d)^2.$$

Since

$$a + b + c + d = 8 - e \quad \text{and} \quad a^2 + b^2 + c^2 + d^2 = 16 - e^2,$$

we find that

$$4(16 - e^2) \geq (8 - e)^2 \quad \text{or} \quad e(5e - 16) \leq 0.$$

Thus, $16/5 \geq e \geq 0$. The upper bound $16/5$ can be obtained by letting $a = b = c = d = 6/5$.

In a similar way, we could find the maximum value of a_n, where

$$a_1 + a_2 + \cdots + a_n = K,$$
$$a_1^2 + a_2^2 + \cdots + a_n^2 = L,$$

and the a_i's are real numbers. For compatibility, it follows by the power mean inequality† that K and L must satisfy $L/n \geq (K/n)^2$. If we have equality, than all the a_i's must be equal. This particular case of the problem has been given many times in various competitions.

I/3. (1980/5).   Prove that, for numbers a, b, c in the interval $[0, 1]$,

$$\frac{a}{b + c + 1} + \frac{b}{(c + a + 1)} + \frac{c}{(a + b + 1)} + (1 - a)(1 - b)(1 - c) \leq 1.$$

Solution.   The function $F(a, b, c)$ on the left hand side of the inequality is convex in each of the variables a, b, c, separately. For say a, two of the terms are linear, while the other two terms are of the form $A/(B + x)$, with $A \geq 0, B > 0$. The graph of the term $A/(B + x)$, for $x > -B$, is a branch of a hyperbola. We can prove its convexity analytically from the easy inequality

$$\frac{A}{B + x} + \frac{A}{B + y} \geq \frac{2A}{B + (x + y)/2},$$

†See Glossary.

or by observing that its second derivative, $2A/(B + x)^3$, is > 0 for $x > -B$. Now we use the elementary but very useful fact that the maximum value of a convex function over an interval must occur at an endpoint of the interval. Thus the maximum value of $F(a, b, c)$ must be achieved at one of the 8 vertices of the cube given by $0 \leqslant a \leqslant 1, 0 \leqslant b \leqslant 1, 0 \leqslant c \leqslant 1$ in rectangular coordinates (a, b, c). In fact, at each vertex, $F(a, b, c) = 1$.

In a similar way, we can establish the more general inequality

$$\text{(1)} \qquad \sum x_i^u / (1 + s - x_i) + \prod (1 - x_i)^v \leqslant 1$$

where $1 \geqslant x_i \geqslant 0$; $u, v \geqslant 1$; and $x_1 + x_2 + \cdots + x_n = s$. This latter inequality is an extension of one due to André Giroux which corresponds to the case $u = v = 1$.

Finally, we give another proof of Giroux's inequality. We can assume without loss of generality that $0 \leqslant x_1 \leqslant x_2 \leqslant \cdots \leqslant x_n \leqslant 1$. Then the left hand side of (1) (with $u = v = 1$) is less than or equal to

$$\sum_1^n \frac{x_i}{1 + s - x_n} + \prod_1^n (1 - x_i)$$

$$= 1 + (x_n - 1)\left\{\frac{1}{1 + s - x_n} - \prod_1^{n-1} (1 - x_i)\right\}.$$

We show that the second term on the right is non-positive. The factor $x_n - 1$ is $\leqslant 0$, so all we need show now is that

$$1 \geqslant (1 + s - x_n) \prod_1^{n-1} (1 - x_i),$$

and this follows from

$$(1 + s - x_n) \prod_1^{n-1} (1 - x_i) \leqslant \prod_1^{n-1} (1 + x_i) \prod_1^{n-1} (1 - x_i)$$

$$= \prod_1^{n-1} (1 - x_i^2) \leqslant 1.$$

I/4. (1977/5). If a, b, c, d, e are positive numbers bounded by p and q, i.e., if they lie in $[p, q]$, $0 < p$, prove that

$$(a + b + c + d + e)\left(\frac{1}{a} + \frac{1}{b} + \frac{1}{c} + \frac{1}{d} + \frac{1}{e}\right) \leqslant 25 + 6\left(\sqrt{\frac{p}{q}} - \sqrt{\frac{q}{p}}\right)^2$$

and determine when there is equality.

Solution. As in I/3, the left hand side $F(a, b, c, d, e)$ of the inequality is a convex function of each of the variables, and consequently the maximum is taken on at one of the 32 vertices of the 5-dimensional cube given by $p \leqslant a, b, c, d, e \leqslant q$. Suppose there are n p's and $5-n$ q's, where $n = 0, 1, \ldots, 5$. Then we have to maximize the quadratic function of n

$$F = (np + \{5 - n\}q)(n/p + \{5 - n\}/q)$$

$$= n^2 + (5 - n)^2 + n(5 - n)\left(\frac{p}{q} + \frac{q}{p}\right)$$

$$= n^2 + (5 - n)^2 + 2n(5 - n) + n(5 - n)\left(\sqrt{p/q} - \sqrt{q/p}\right)^2$$

$$= 25 + n(5 - n)\left(\sqrt{p/q} - \sqrt{q/p}\right)^2.$$

Finally, F takes on its maximum value $25 + 6(\sqrt{p/q} - \sqrt{q/p})^2$ for $n = 2$ or 3. So, for equality, two or three of a, b, c, d, e are p and the rest are q.

Again, as in the previous problems, the inequality can be generalized. Proceeding in the same way as before, with m variables a_i satisfying $0 < p \leqslant a_1, a_2, \ldots, a_m \leqslant q$, we obtain

$$\sum a_i \sum \frac{1}{a_i} \leqslant m^2 + \left[\frac{m}{2}\right]\left(m - \left[\frac{m}{2}\right]\right)\left(\sqrt{\frac{p}{q}} - \sqrt{\frac{q}{p}}\right)^2.$$

I/5. (1980/2). Determine the maximum number of different three-term arithmetic progressions which can be chosen from a sequence of n real numbers

$$a_1 < a_2 < \cdots < a_n.$$

Solution. Consider the three-term arithmetic progressions having the middle term a_i for $1 < i < n$. For $n(\text{odd}) = 2k + 1$ and middle term a_i with $1 < i \leqslant k$, there are the $i - 1$ possible first terms $a_1, a_2, \ldots, a_{i-1}$. For all such a_i, the number of such progressions is at most $1 + 2 + \cdots + (k - 1) = k(k - 1)/2$. As for possible third terms when the middle term is a_i with $k < i < n$, there are the $n - i$ terms $a_{i+1}, a_{i+2}, \ldots, a_n$. The number of such progressions is at most $1 + 2 + \cdots + k = k(k + 1)/2$. Hence the total number of three-term progressions, for n odd, is at most

$$\frac{k(k - 1)}{2} + \frac{k(k + 1)}{2} = k^2 = \frac{(n - 1)^2}{4}.$$

For n even, we proceed in a similar way and find that the number of progressions is at most

$$\frac{n^2 - 2n}{4}.$$

Both results for n even or odd can be given by the single expression $[(n-1)^2/4]$, where $[x]$ is the greatest integer not exceeding x. Also, it is easy to see that these bounds can be realized when $a_1, a_2, \ldots, a_n$ is itself an arithmetic progression.

I/6. (1972/4). Let R denote a non-negative rational number. Determine a fixed set of integers a, b, c, d, e, f, such that for *every* choice of R,

$$\left| \frac{aR^2 + bR + c}{dR^2 + eR + f} - \sqrt[3]{2} \right| < \left| R - \sqrt[3]{2} \right|.$$

Solution. The inequality says that no matter what the rational number R is, in particular, no matter how close or how far it is to the irrational number $2^{1/3}$, the rational number

$$f(R) = \frac{aR^2 + bR + c}{dR^2 + eR + f}$$

is even closer to $2^{1/3}$. Thus we can use $f(R)$ to generate closer and closer approximations to $2^{1/3}$.

As $R \to 2^{1/3}$ through a sequence of non-negative numbers, the right side of the inequality approaches 0. Consequently the left side must vanish if we set $R = 2^{1/3}$, so

(1) $\qquad a \cdot 2^{2/3} + b \cdot 2^{1/3} + c = 2d + e \cdot 2^{2/3} + f \cdot 2^{1/3}.$

For equation (1) to hold, we must have

$$a = e, \quad b = f, \quad c = 2d.$$

On substituting back in the inequality, we find a common factor $R - 2^{1/3}$ which we cancel out to obtain

(2) $\qquad \left| aR + b - d \cdot 2^{1/3}(R + 2^{1/3}) \right| \leqslant \left| dR^2 + aR + b \right|.$

To satisfy (2) it suffices to let a, b, d be positive integers and make the quantity on the left, within the absolute signs, positive. Just let $a > d \cdot 2^{1/3}$ and $b > d \cdot 2^{2/3}$. Two simple choices are

$$d = 1, \, a = b = 2 \quad \text{and} \quad d = 3, \, a = 4, \, b = 5.$$

These lead, respectively, to the *always* better rational approximations

(3) $\qquad f_1(R) = \dfrac{2R^2 + 2R + 2}{R^2 + 2R + 2}, \quad f_2(R) = \dfrac{4R^2 + 5R + 6}{3R^2 + 4R + 5}$

to $2^{1/3}$ than R.

If we started, for example, with $R_0 = 5/4$, then the first formula gives

$$f_1(R_0) = f_1\left(\tfrac{5}{4}\right) = \frac{122}{97} \quad \text{with } \left| f_1(R_0) - 2^{1/3} \right| \approx .00219$$

and the second

$$f_2(R_0) = f_2\left(\tfrac{5}{4}\right) = \frac{296}{235} \quad \text{with } \left| f_2(R_0) - 2^{1/3} \right| \approx .00035.$$

In a similar fashion, one can *always* obtain better rational approximations than R to $N^{1/r}$, where N, r are positive integers.

If we iterate (repeatedly use) the above formulae, we obtain a sequence of approximations converging to $2^{1/3}$. There are much more efficient schemes for such approximations, e.g. the Newton algorithm [65, pp. 174–179] and continued fractions [see N.M.L. #9 or A. N. Khovanskii, *The Application of Continued Fractions and their Generalization to Problems in Approximation*, Noordhoff, Groningen, 1963, pp. 194–202]. One Newton formula for approximating $N^{1/r}$ is

$$R_1 = R_0 - \frac{(R_0^r - N)}{rR_0^{r-1}}.$$

For our case $N = 2$, $r = 3$, and $R_0 = \tfrac{5}{4}$, we get

$$R_1 = 63/50 \quad \text{and} \quad \left| R_1 - 2^{1/3} \right| \approx .000079.$$

Note that in using the Newton scheme here, the second approximant may be farther from the root than the first; nevertheless the sequence will still converge to $2^{1/3}$. Our formulae (3) *always* yield a next approximant closer to the desired root than the previous approximant.

For other equations, the Newton sequence can converge to an unwanted result, or it can diverge if the first approximant is too far away from the root of interest. Incidentally, one can give a pathological case [see Math. Intelligencer (1984, No. 3), pp. 28, 37; (1985, No. 3), p. 36] of the Newton iteration in which the scheme will never converge regardless of how good the first approximation is. Also, in approximating the solution of simultaneous non-linear equations with many variables, the difficult part is to find a sufficiently good first approximation.

I/7. (1985/2). Determine each real root of

$$x^4 - (2 \cdot 10^{10} + 1)x^2 - x + 10^{20} + 10^{10} - 1 = 0$$

correct to four decimal places.

Solution (by Bill Cross). By Descartes' rule of signs, the number of positive roots is at most two. We rewrite the equation in the form

(1) $$\left(x^2 - 10^{10} - \tfrac{1}{2} \right)^2 = x + \tfrac{5}{4}.$$

The left side is always non-negative while the right side, for $x < 0$, is non-negative only for $-\frac{5}{4} \leqslant x < 0$. In this interval, the left side is approximately 10^{20} while the right side is less than $\frac{5}{4}$. Hence (1) has no negative roots.

To find the positive roots, we use successive approximation. Since x^2 dominates x, and 10^{10} dominates $\frac{1}{2}$ and $\frac{5}{4}$, we approximate (1) by the simple quadratic equation in x^2

$$(x^2 - 10^{10})^2 = 0,$$

from which we obtain $x = 10^5$ as first approximation to a positive zero of the given polynomial. Now replace x in the right side of Equation (1) by 10^5; still omitting the $\frac{1}{2}$ and $\frac{5}{4}$, we obtain

$$(x^2 - 10^{10})^2 = 10^5 \text{ satisfied by } x = \left(10^{10} + \sqrt{10^5}\right)^{1/2},$$

which is approximately

$$10^5 \pm \frac{\sqrt{10^5}}{2 \cdot 10^5} = 10^5 \pm \frac{5\sqrt{10}}{10^4} \approx 10^5 + 15.8 \cdot 10^{-4} \approx 10^5 \pm .0016.$$

[Here we used the fact that when $a \gg b$, then $a + b/2a = R$ differs from $(a^2 + b)^{1/2} = B$ by less than $b^2/8a^3$, a very small quantity. Indeed,

$$A^2 = a^2 + b + b^2/4a^2,$$

so that

$$A^2 - B^2 = (A - B)(A + B) = b^2/4a^2,$$

and

$$A - B = \frac{b^2}{4a^2(A + B)} < \frac{b^2}{4a^2(2a)}.$$

In our case $a = 10^5, b = 10^{5/2}$ so that $A - B < 0.125(10)^{-10}$.]

We conjecture that $x = 10^5 \pm .0016$ are the zeros of the function

$$f(x) = \left(x^2 - 10^{10} - \frac{1}{2}\right)^2 - x - \frac{5}{4},$$

correct to four decimal places.

For verification, we compute

$$f(10^5 \pm .00155)$$

$$= \left(\pm 310 + (.00155)^2 - \frac{1}{2}\right)^2 - (10^5 \pm .00155) - 1.25(311)^2 - 10^5$$

$$< 0.$$

On the other hand,

$$f(10^5 \pm .00165)$$

$$= \left(\pm 330 + (.00165)^2 - \frac{1}{2}\right)^2 - (10^5 \pm .00165) - 1.25$$

$$> 320^2 - 10^5 - 2 > 0.$$

The proof is complete.

Second Solution. More generally, we consider the equation

$$x^4 - (2a^2 + 1)x^2 - x + a^4 + a^2 - 1 = 0,$$

where a is a very large number compared to 1. By Descartes' rule of signs, the number of positive roots is at most two. The equation may be rewritten as $P(x) \equiv (x^2 - a^2)^2 - (x^2 - a^2) - x - 1 = 0$. Since $P(a - 1) > 0$, $P(a) > 0$ and $P(a + 1) > 0$, there are two positive roots lying in the intervals $(a - 1, a)$ and $(a, a + 1)$, respectively. Let $a + e$ be a first approximation to a positive root r. Substituting into the equation, we have

$$4a^2 e^2 - a \approx 0 \quad \text{or} \quad e \approx \pm \frac{1}{2\sqrt{a}}.$$

It is easy to verify that

$$P\left(a - \frac{1}{\sqrt{a}}\right) > 0, \quad P\left(a - \frac{1}{2\sqrt{a}}\right) < 0, \quad \text{and } P\left(a + \frac{1}{2\sqrt{a}}\right) < 0, \quad P\left(a + \frac{1}{\sqrt{a}}\right) > 0.$$

Hence the positive roots lie in the intervals

$$\left(a - \frac{1}{\sqrt{a}}, a - \frac{1}{2\sqrt{a}}\right) \quad \text{and} \quad \left(a + \frac{1}{2\sqrt{a}}, a + \frac{1}{\sqrt{a}}\right).$$

For greater accuracy, we write

$$r = a + \frac{1}{2\sqrt{a}} + f.$$

Substituting into the equation, we have $4a\sqrt{a}f \mp \sqrt{a} \approx 0$ or $f \approx 1/4a$. It is easy to verify that

$$P\left(a - \frac{1}{2\sqrt{a}} + \frac{1}{4a}\right) \quad \text{and} \quad P\left(a + \frac{1}{2\sqrt{a}} + \frac{1}{4a}\right)$$

are both negative. Let $\varepsilon > 0$ be a very small number compared to 1. It is straightforward to verify that

$$P\left(a - \frac{1}{2\sqrt{a}} + \frac{1}{(4 + \varepsilon)a}\right) \quad \text{and} \quad P\left(a + \frac{1}{2\sqrt{a}} + \frac{1}{(4 - \varepsilon)a}\right)$$

are positive. Hence the roots lie in

$$\left(a - \frac{1}{2\sqrt{a}} + \frac{1}{(4 + \varepsilon)a}, \quad a - \frac{1}{2\sqrt{a}} + \frac{1}{4a}\right)$$

and

$$\left(a + \frac{1}{2\sqrt{a}} + \frac{1}{4a}, \quad a + \frac{1}{2\sqrt{a}} + \frac{1}{(4 - \varepsilon)a}\right).$$

Note that $(3.1)^2 < 10 < (3.2)^2$. We have, for $a = 10^5$,

$$a - \frac{1}{2\sqrt{a}} + \frac{1}{(4 + \varepsilon)a} > 10^5 - \frac{3.2}{2000}, \quad \text{while} \quad a - \frac{1}{2\sqrt{a}} + \frac{1}{4a} < 10^5 - \frac{3.1}{2000}.$$

Hence the smaller positive root is equal to 99999.9984 correct to four decimal places. Similarly, we have

$$a + \frac{1}{2\sqrt{a}} + \frac{1}{4a} > 10^5 + \frac{3.1}{2000}, \quad \text{while} \quad a + \frac{1}{2\sqrt{a}} + \frac{1}{(4 - \varepsilon)a} < 10^5 + \frac{3.2}{2000}.$$

Hence the larger positive root is equal to 100000.0016 correct to four decimal places. To complete the argument, we show that $P(x) = 0$ has no negative roots. Letting $y = -x > 0$, the equation becomes $(y^2 - a^2)^2 + y - 1 = y^2 - a^2$. If it had a positive root $y = r$, clearly $r > a$. Moreover, we have $r^2 - a^2 > (r^2 - a^2)^2$ and $r^2 - a^2 > r - 1$. From these two inequalities, we have that $1 + a^2 > r$ and $r > \frac{1}{2}(1 + \sqrt{4a^2 - 3})$. This is a contradiction since $1 + a^2 < (1 + \sqrt{4a^2 - 3})^2/4$.

I/8. (1974/4). A father, mother and son hold a family tournament, playing a two person board game with no ties. The tournament rules are:

(i) The weakest player chooses the first two contestants.

(ii) The winner of any game plays the next game against the person left out.

(iii) The first person to win two games wins the tournament.

The father is the weakest player, the son the strongest, and it is assumed that any player's probability of winning an individual game from another player does not change during the tournament.

Prove that the father's optimal strategy for winning the tournament is to play the first game with his wife.

Solution. Let F, M and S denote father, mother and son, respectively, and let $W > L$ abbreviate "player W wins and player L loses" in an individual game.

If F and M play the first game, then F can win the tournament in any one of the following three mutually exclusive game sequences:

(1) $F > M, \quad F > S;$

(2) $F > M, \quad S > F, \quad M > S, \quad F > M;$

(3) $M > F, \quad S > M, \quad F > S, \quad F > M;$

(Note that even if the weakest player wins, he does it within at most four games.)

If F and S play the first game, the winning sequences are the same as the three above with M and S interchanged. If M and S play the first game, then F can win in the two sequences

(4) $S > M, \quad F > S, \quad F > M;$

(5) $M > S, \quad F > M, \quad F > S;$

observe that (5) is the same as (4) with M and S interchanged.

Let $\text{Prob}(W > L) = \overline{WL}$; since there are no ties, we have $\overline{WL} + \overline{LW} = 1$.

If F and M play the first game, the probability that F wins the tournament is

$$P_{FM} = \overline{FM} \cdot \overline{FS} + \overline{FM} \cdot \overline{SF} \cdot \overline{MS} \cdot \overline{FM} + \overline{MF} \cdot \overline{SM} \cdot \overline{FS} \cdot \overline{FM}.$$

If F and S play the first game, then F's probability of winning the tournament is

$$P_{FS} = \overline{FS} \cdot \overline{FM} + \overline{FS} \cdot \overline{MF} \cdot \overline{SM} \cdot \overline{FS} + \overline{SF} \cdot \overline{MS} \cdot \overline{FM} \cdot \overline{FS}.$$

And if M and S play the first game, it is

$$P_{MS} = \overline{SM} \cdot \overline{FS} \cdot \overline{FM} + \overline{MS} \cdot \overline{FM} \cdot \overline{FS}$$

$$= (\overline{SM} + \overline{MS})\overline{FS} \cdot \overline{FM}$$

$$= \overline{FS} \cdot \overline{FM}.$$

Since P_{MS} is clearly less than either P_{FM} or P_{FS}, we need only compare the latter two to find F's best strategy.

$$P_{FM} - P_{FS} = (\overline{FM} - \overline{FS})(\overline{SF} \cdot \overline{MS} \cdot \overline{FM} + \overline{MF} \cdot \overline{SM} \cdot \overline{FS}).$$

Since S is the strongest player, it is tacitly assumed that $\overline{FM} > \overline{FS}$, and so $P_{FM} > P_{FS}$.

For a related more difficult problem, see Problem 74-23, SIAM Review, vol. 16 (1974), p. 547.

1/9. (1979/3). Given three identical n-faced dice whose corresponding faces are identically numbered with arbitrary integers. Prove that if they are tossed at random, the probability that the sum of the bottom three face numbers is divisible by three is greater than or equal to $1/4$.

Solution. We may replace the numbers on the faces of each die by their residues modulo 3. Letting x, y, z be the probabilities, respectively, that the numbers $0, 1, 2 \pmod 3$ come down on each die, we first want an expression for the probability P that the sum of the numbers on the three bottom faces of the three dice is $0 \pmod 3$. This event happens only for the throws $(0, 0, 0), (1, 1, 1), (2, 2, 2)$, and the six permutations of $(0, 1, 2)$. Thus the desired probability is given by

$$P = x^3 + y^3 + z^3 + 6xyz,$$

where $x + y + z = 1$ and $x, y, z \geq 0$.

We now show in several ways that $P \geq 1/4$.

1. $P = (x^3 + y^3 + z^3 - 3xyz) + 9xyz$

$$= (x + y + z)\{(x + y + z)^2 - 3(yz + zx + xy)\} + 9xyz$$

$$= 1 - 3(yz + zx + xy) + 9xyz.$$

Then $P \geq 1/4$ is equivalent to $1 + 12xyz \geq 4(yz + zx + xy)$. We can assume without loss of generality that $x \geq y \geq z$ so that $x \geq 1/3$. Then the last inequality can be written

$$1 + 4yz(3x - 1) \geq 4(zx + xy) = 4x(1 - x).$$

Since $3x - 1 \geq 0$ and $4x(1 - x) \leq 1$, $P \geq 1/4$. There is equality *iff* one of x, y, z is zero and the other two are equal.

2. We convert our inequality $x^3 + y^3 + z^3 + 6xyz \geq \frac{1}{4}$, with $x + y + z = 1$ to the equivalent homogeneous form

$$4(x^3 + y^3 + z^3 + 6xyz) \geq (x + y + z)^3.$$

Expanding out, we get

$$\sum x^3 + 6xyz \geq \sum x^2 y,$$

where the summations are symmetric over x, y, z. This is a weaker inequality than Schur's inequality [120; p. 119], i.e.,

$$\sum x(x - y)(x - z) = \sum x^3 + 3xyz - \sum x^2 y \geq 0,$$

and thus follows from it.

3. The standard calculus method of minimizing the function

$$P(x, y, z) = x^3 + y^3 + z^3 + 6xyz$$

constrained to the portion of the plane $x + y + z = 1$ in the positive octant is straightforward.

We first show that P has no minimum in the interior of the region in question. According to the Lagrange multiplier method, at such a point the gradient of P would have to be parallel of the gradient of the constraint function. Now

$$\text{grad } P = (3x^2 + 6yz, 3y^2 + 6xz, 3z^2 + 6xy),$$

and

$$\text{grad}(x + y + z) = (1, 1, 1),$$

so

$$x^2 + 2yz = a, \quad y^2 + 2xz = a, \quad z^2 + 2xy = a,$$

where a is a parameter. On subtracting the latter equations from one another, we get $(x - y)(x + y - 2z) = 0 = (x - y)(1 - 3z)$, etc. Thus the only interior critical point is $x = y = z = 1/3$, yielding $P(x, y, z) = 1/3$. Any other critical point must be on the boundary, where at least one of x, y, z, say z is 0; there $P(x, y, z) = x^3 + (1 - x)^3 = 1 - 3x(1 - x)$. This quadratic function has its minimum at $x = 1/2$ giving $P(\frac{1}{2}, \frac{1}{2}, 0) = \frac{1}{4}$. Also $P(\frac{1}{2}, 0, \frac{1}{2}) = P(0, \frac{1}{2}, \frac{1}{2}) = \frac{1}{4}$. The remaining critical points are

$(1, 0, 0), (0, 1, 0), (0, 0, 1)$, giving $P(x, y, z) = 1$, clearly the global maximum. Thus the global minimum is $\frac{1}{4}$.

For a more general unsolved problem, see Problem 80-5, SIAM Review, vol. 22 (1980), p. 99.

I/10. (1981/5). If x is a positive real number, and n is a positive integer, prove that

$$[nx] > \frac{[x]}{1} + \frac{[2x]}{2} + \frac{[3x]}{3} + \cdots + \frac{[nx]}{n},$$

where $[t]$ denotes the greatest integer less than or equal to t. For example, $[\pi] = 3$ and $[\sqrt{2}] = 1$.

First Solution. Since for any integer k, $[k + x] = k + [x]$, it suffices to prove the inequality for $0 < x < 1$. Observe that both members of the inequality are piecewise constant increasing functions of x. The one on the right increases only at rational points $x = p/q$, where $(p, q) = 1$, and $2 \leqslant q \leqslant n$ and $1 \leqslant p \leqslant q - 1$. It is therefore enough to prove that the inequality holds at these points. For $k = 1, 2, \ldots, n$, let integers a_k and b_k be defined by

$$kp = a_k q + b_k, \qquad 0 \leqslant b_k < q.$$

The inequality to be proved now reads

$$a_n \geqslant a_1 + a_2/2 + a_3/3 + \cdots + a_n/n.$$

We claim that, since $(p, q) = 1$, the numbers $b_1, b_2, \ldots, b_{q-1}$ are a rearrangement of $1, 2, \ldots, q - 1$. For, suppose two of them, say b_i and b_j satisfied

$$b_i \equiv b_j \pmod{q}, \qquad ip = a_i q + b_i, \qquad jp = a_j q + b_j;$$

subtracting, we get $(i - j)p = mq$, m an integer. According to Euclid's lemma, q divides $i - j$; since $0 < i, j < q$, this is possible only if $i = j$.

By the rearrangement inequality [119, p. 261],

$$b_1 + b_2/2 + b_3/3 + \cdots + b_{q-1}/(q - 1) \geqslant q - 1.$$

Consequently,

$$a_n + (q - 1)/q \geqslant a_n + b_n/q = np/q$$

$$= a_1 + \frac{a_2}{2} + \cdots + \frac{a_n}{n} + \frac{1}{q}\left(b_1 + \frac{b_2}{2} + \cdots + \frac{b_n}{n}\right)$$

$$\geqslant a_1 + \frac{a_2}{2} + \cdots + \frac{a_n}{n} + (q - 1)/q.$$

Hence the desired result follows.

Second solution. Note that

$$nx = x + \frac{2x}{2} + \frac{3x}{3} + \cdots + \frac{nx}{n},$$

and if

$$[nx] \geq [x] + \frac{[2x]}{2} + \frac{[3x]}{3} + \cdots + \frac{[nx]}{n},$$

then

$$nx - [nx] \leq (x - [x]) + \tfrac{1}{2}(2x - [2x]) + \cdots + \frac{1}{n}(nx - [nx]).$$

Let $\{kx\}$ be the fractional part of kx, i.e. $kx - [kx]$. Since $nx = x + 2x/2 + 3x/3 + \cdots + nx/n$, we must prove that

$$\{nx\} \leq \{x\} + \frac{\{2x\}}{2} + \frac{\{3x\}}{3} + \cdots + \frac{\{nx\}}{n}.$$

Both sides of this inequality are piecewise linear functions of x with slope n. All discontinuities are jumps down. The left side jumps at $\frac{1}{n}, \frac{2}{n}, \frac{3}{n}, \cdots$. So, if the inequality is violated for any x, it is violated up to the next multiple of $1/n$. Let the next multiple be k/n, and $x = (k/n) - \varepsilon$, where $0 < \varepsilon < 1/n$. Let $(n, k) = d$ and $n/d = c$. Modulo n, the numbers $k, 2k, \ldots, ck$ are permutations of $0, d, 2d, \ldots, (c-1)d$. Hence $\{x\}, \{2x\}/2, \ldots, \{cx\}/c$ are within ε of some rearrangement of $d/n, 2d/n, \ldots, cd/n$. Therefore

$$\{x\} + \frac{\{2x\}}{2} + \cdots + \frac{\{cx\}}{c} \geq \frac{d}{n} + \frac{2d}{2n} + \cdots + \frac{cd}{cn} - c\varepsilon$$

$$\geq \frac{cd}{n} - n\varepsilon = 1 - n\varepsilon = \{k - n\varepsilon\} = \{nx\}$$

Note. The first solution of this problem is essentially due to Professor A.M. Gleason. The second solution is due to Professor Peter Ungar.

Combinatorics and Probability

C. & P./1. (1976/1).

(a) Suppose that each square of a 4×7 chessboard, as shown above, is colored either black or white. Prove that with *any* such coloring, the board

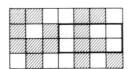

must contain a rectangle (formed by the horizontal and vertical lines of the board such as the one outlined in the figure) whose four distinct unit corner squares are all of the same color.

(b) Exhibit a black-white coloring of a 4×6 board in which the four corner squares of every rectangle, as described above, are not all of the same color.

Solution. (a) We prove a stronger version by showing that the stated result is even valid for a 3×7 board. We call any column (consisting of 3 unit squares) *black* if it has more black unit squares than white ones; otherwise we call it *white*. Since there are seven columns, at least four of them are of the same kind, say *black*. We now show that there is a rectangle whose vertices are in these four *black* columns and whose corner squares are all black. We can even assume that each of these four *black* columns has one white unit square. Since the white square can be in one of only three positions, it follows that two of these four columns are identically colored giving the desired result.

(b) For the 4×6 chessboard, there are $\binom{4}{2} = 6$ different arrangements of two B's and two W's in a column. One such coloring is given in the figure below; it contains no rectangle with identically colored corner squares.

B	B	B	W	W	W
B	W	W	W	B	B
W	B	W	B	W	B
W	W	B	B	B	W

General results: The stated result in (a) is not valid for a $2 \times n$ board (all n), a $3 \times n$ board ($n < 7$), a $4 \times n$ board ($n < 7$), and a $5 \times n$ board ($n < 5$). The $2 \times n$ case is settled by the following counterexample to (a):

$$\begin{matrix} B & B & B & B & \cdots \\ W & W & W & W & \cdots \end{matrix}$$

The remaining cases follow from (b) above. Finally, the stated result in (a) is valid for all $5 \times n$ boards with $n \geqslant 5$. To prove this it suffices to take $n = 5$. As before we call any column (consisting of 5 unit squares) *black* if it has more black unit squares than white ones; otherwise we call it *white*. Since there are five columns, at least three of them are of the same kind, say *black*. We now show that there is a rectangle whose vertices are in these three *black* columns and whose corner squares are all black.

In this 5×3 subboard, there are at most six white squares. Suppose one row consists of only black squares. Then at least one other row will have at least two black squares, yielding the desired rectangle. On the other hand, if there are no rows consisting of only black squares, then each row has at least one white square. Hence there are at least four rows each with exactly one white square. As in (a), two of these rows must be identically colored, giving the desired result.

C. & P./2. (1979/5).   Nine mathematicians meet at an international conference and discover that among any three of them, at least two speak a common language. If each of the mathematicians can speak at most three languages, prove that there are at least three of the mathematicians who can speak the same language.

Solution.   Our proof is indirect. We assume that at most two mathematicians speak a common language. Each mathematician can speak to at most three others, one for each language he or she knows. Suppose mathematician M_1 can only speak with M_2, M_3, and M_4. Now mathematician M_5 can speak with at most three of M_2, M_3, and M_4 or at most three of M_6, M_7, M_8, and M_g. This leaves one of the last four who cannot speak with M_1 or M_5 giving the desired contradiction.

The above result has been generalized as follows:
Determine the largest integer $N(t, m, p)$, with $t \geqslant 1, m \geqslant 2, p \geqslant 3$, such that there exists a set of N persons satisfying the three conditions:
  (i) Each person speaks at most t languages,
  (ii) Among any m persons, two speak a common language,
  (iii) No p persons speak a common language.
The given problem corresponds to showing that $N(3,3,3) < 9$. Actually, $N(3,3,3) = 8$. For a proof, let M_1, M_2, M_3, and M_4 be such that each pair can communicate in a unique language. Also let M_5, M_6, M_7, and M_8 form a second set with the same property. Then each mathematician speaks exactly 3 languages and each language is spoken by exactly 2 mathematicians. Finally, among any 3 mathematicians, 2 of them will belong to the same set by the Pigeonhole Principle, and hence can communicate.
Among further results obtained in a series of papers [1, 2, 3, 4] are that

$$N(1, m, p) = (m - 1)(p - 1),$$

$$N(t, m, 3) = (m - 1)(t + 1),$$

$$N(2, m, p) = 3(m - 1)(p - 1)/2 \quad \text{if } p \equiv 1 (\mathrm{mod}\, 2),$$

$$= (m - 1)(3p - 4)/2 \quad \text{if } p \equiv 0 (\mathrm{mod}\, 2),$$

$$N(t, m, 4) = (m - 1)(2t + 1) \qquad \text{if } t \equiv 0, 1 (\mathrm{mod}\, 3),$$

$$= 2(m - 1)t \qquad \text{if } t \equiv 2 (\mathrm{mod}\, 3).$$

Also, various bounds are obtained for other cases.

References

1. H. L. Abbott, D. Hanson, A. C. Liu, An extremal problem in graph theory, Quart. J. Math. Oxford Ser. 31 (1980) pp. 1–7.
2. H. L. Abbott, M. Katchalski, A. C. Liu, An extremal problem in graph theory II, J. Austral Math. Soc. (Series A), 29 (1980) pp. 417–424.
3. H. L. Abbott, M. Katchalski, A. C. Liu, An extremal problem in hypergraph theory, Discrete Math. Anal. and Comb. Comp., Conference proceedings, School of Computer Science, Univerity of New Brunswick, Fredericton, 1980, pp. 74–82.
4. H. L. Abbott, D. Hanson, A. C. Liu, An extremal problem in hypergraph theory II, J. Austral. Math. Soc. (Series A), 31 (1981) pp. 129–135.

C. & P./3. (1982/1).   In a party with 1982 persons, among any group of four there is at least one person who knows each of the other three. What is the minimum number of people in the party who know everyone else?

Solution.   *Case* 1. We assume "knowing" is not a symmetrical relation, i.e., A may know B but B does not know A. For this case no one need know everyone else. Just consider all the people arranged in a circle such that each person knows everyone else except the person next to him clockwise.

Case 2. We assume "knowing" is a symmetrical relation. Let $\{P_1, P_2\}$ be a pair who do not know each other. Then if $\{P_3, P_4\}$ is a pair disjoint from the first pair, P_3 and P_4 must know each other since one of P_1, P_2, P_3, P_4 knows the other three by hypothesis. So if there is a third person P_3 who does not know everyone, it must be P_1 or P_2 he(she) does not know. If there were a fourth person P_4 who did not know everyone it would again be P_1 or P_2 he(she) did not know, but then $\{P_1, P_2, P_3, P_4\}$ would violate the hypothesis. Thus all except at most three people must know everyone else.

C. & P./4. (1985/4).   There are n people at a party. Prove that there are two people such that, of the remaining $n - 2$ people, there are at least $\lfloor n/2 \rfloor - 1$ of them, each of whom knows both or else knows neither of the two. Assume that "knowing" is a symmetrical relation; $\lfloor x \rfloor$ denotes the greatest integer less than or equal x.

Solution.   Given two people at the party, we describe a third person as "mixed" with respect to that pair, if that person knows exactly one of the two. Thus a person who knows exactly k people at the party is mixed with respect to $k(n - 1 - k)$ pairs. By the A.M.-G.M. inequality, each person is mixed with respect to at most $(n - 1)^2/4$ pairs. Thus there are at most

$$\frac{n(n-1)^2}{4} = \frac{n-1}{2}\binom{n}{2}$$

person-mixed pair combinations, and so for at least one of the $\binom{n}{2}$ pairs at most $\lfloor (n-1)/2 \rfloor$ of the remaining people are mixed. For this pair, there are at least

$$n - 2 - \left\lfloor \frac{n-1}{2} \right\rfloor = \left\lfloor \frac{n}{2} \right\rfloor - 1$$

others who know either both or neither of the two.

C. & P./5. (1986/2).   During a certain lecture, each of five mathematicians fell asleep exactly twice. For each pair of these mathematicians, there was some moment when both were sleeping simultaneously. Prove that, at some moment, some three were sleeping simultaneously.

Solution.   Our proof is indirect. Assuming that no three mathematicians ever slept simultaneously, we find there were 10 non-overlapping time *intervals*, one *interval* of common dozing for each of the $\binom{5}{2}$ pairs of mathematicians. Each such *interval* was initiated by a *moment* when one of the mathematicians in the pair fell asleep. Since each of the 5 mathematicians fell asleep twice, there were precisely 10 such *moments* so that each initiated a different *interval*. However, since two of the mathematicians were sleeping during the first *interval*, two of the 10 *moments* had already occurred, leaving 8 *moments* to initiate the remaining 9 *intervals*. This is impossible and the desired result follows from this contradiction.

This problem as well as the subsequent problem C. & P./8 are special cases of results given in the paper by M. Katchalski and A. Liu, *Arcs on the circle and p-tuplets on the line*, Discrete Math., 27 (1979) pp. 59–69.

C. & P./6. (1979/5).   A certain organization has n members, and it has $n + 1$ three-member committees, no two of which have identical membership. Prove that there are two committees which share exactly one member.

Solution.   Clearly, the problem statement implies that $n \geqslant 5$. Our proof is indirect, so we assume that for any pair of committees, they either have two common members or else they are disjoint (no common members).

Since each of the $n + 1$ committees has three members, there are $3n + 3$ committee positions. If each of the n persons served on at most 3 committees, at most $3n$ positions could be filled. Therefore one person must belong to at least four committees. Let A denote one such person, and let $\mathscr{C}_1, \mathscr{C}_2, \mathscr{C}_3, \mathscr{C}_4$ denote four committees to which he belongs. Let $\{A, B, C\}$ be the membership of $\mathscr{C}_1$; since by our assumption two committees that have A in common must have another member in common, either B or C belongs to two of the remaining committees $\mathscr{C}_2, \mathscr{C}_3, \mathscr{C}_4$. Thus we may assume the following committee membership form:

$$\mathscr{C}_1: A, B, C \quad \mathscr{C}_2: A, B, D \quad \mathscr{C}_3: A, B, E \quad \mathscr{C}_4: A, \ldots.$$

Now $\mathscr{C}_4$ must also contain B, for otherwise it cannot share two members with each of the committees $\mathscr{C}_1, \mathscr{C}_2, \mathscr{C}_3$. Similarly, any additional committee $\mathscr{C}_5$ to which A belongs also contains B. By symmetry, any additional committee that contains B also contains A. Denote by $\mathscr{C}_1, \mathscr{C}_2, \ldots, \mathscr{C}_k$ all committees that contain A or B; by the above argument, they contain A and B. This accounts for k committees with a total membership of $k + 2$ persons. The remaining $n - k - 2 = m$ persons serve on the $n + 1 - k = m + 3$ remaining committees $\mathscr{C}_{k+1}, \mathscr{C}_{k+2}, \ldots, \mathscr{C}_{n+1}$, none of which contain people who serve on any of the first k committees.

Thus we have shown that, if the statement to be proved is false for n, then it is false for some $m < n$. It follows that the statement is true for all n, otherwise there would be a smallest number of persons for which it is false. But our argument shows that then there would be a still smaller number for which it is false.

C. & P./7. (1981/2). Every pair of communities in a county are linked directly by exactly one mode of transportation: bus, train or airplane. All three modes of transportation are used in the county with no community being serviced by all three modes and no three communities being linked pairwise by the same mode. Determine the maximum number of communities in this county.

Solution. We will show that the maximum number is four and this can be achieved by having communities A and B linked by train, C and D by bus and the remaining pairs by airplane.

First we show that no community can be linked to three others by the same mode of transportation. The proof is indirect. Assume A is linked to $B, C,$ and D by train. Now $B, C,$ and D cannot be linked pairwise by train, nor can all three be linked to one another only by bus or only by airplane. It follows that one of $B, C,$ and D is serviced by all three modes which gives the contradiction. Also, it follows that each community is serviced by two modes of transportation with each mode linking it to two other communities. This limits the number of communities to at most five.

We now assume that there are five communities and obtain a contradiction giving the desired result. Suppose that A is linked to B and C by mode M_1 and to D and E by M_2. Since two M_1 modes leave C, we may assume without loss of generality that C is linked to D by M_1. Then since the link between D and E cannot be M_2 nor M_3, it must be M_1. Then the link between C and E must be M_2. Continuing in this manner, mode M_3 cannot be used which gives the contradiction.

C. & P./8. (1983/3).   Each set of a finite family of subsets of a line is a union of two closed intervals. Moreover, any three of the sets of the family have a point in common. Prove that there is a point which is common to at least half the sets of the family.

Solution.   Let the family be given by $\{ F_i \colon 1 \leqslant i \leqslant n \}$. The union of two closed intervals can always be written $F_i = [a_i, b_i] \cup [c_i, d_i]$, where $a_i \leqslant b_i \leqslant c_i \leqslant d_i$. Let $a = \max\{a_i \colon 1 \leqslant i \leqslant n\}$ and $d = \min\{d_i \colon 1 \leqslant i \leqslant n\}$. Now $a = a_j$ for some j and $d = d_k$ for some k. We claim that every F_i contains a_j or d_k, from which our result follows. On the contrary, suppose that some F_i contains neither a_j nor d_k. Since $a_i \leqslant a_j$, we must have $F_j \cap [a_i, b_i] = \varnothing$. Since $d_i \geqslant d_k$, we must have $F_k \cap [c_i, d_i] = \varnothing$. It follows that $F_i \cap F_j \cap F_k = \varnothing$ which contradicts the hypothesis.

C. & P./9. (1985/5).   Let $a_1, a_2, a_3, \cdots$ be a non-decreasing sequence of positive integers. For $m \geqslant 1$, define $b_m = \min\{n \colon a_n \geqslant m\}$, that is, b_m is the minimum value of n such that $a_n \geqslant m$. If $a_{19} = 85$, determine the maximum value of

$$a_1 + a_2 + \cdots + a_{19} + b_1 + b_2 + \cdots + b_{85}.$$

Solution.   More generally, we will show that if $a_q = p$, then

$$S_{p,q} \equiv a_1 + a_2 + \cdots + a_q + b_1 + b_2 + \cdots + b_p = p(q + 1).$$

In particular, for the case $q = 19$, $p = 85$, we have the sum $p(q + 1) = 1700$.

If $a_i = p$ for all $1 \leqslant i \leqslant q$, then $b_j = 1$ for all $1 \leqslant j \leqslant p$. Hence, $S_{p,q} = qp + p$ as required. If not, let t be the largest index such that $a_t < p$. Let $a_t = u$. If a_t is increased by 1, then all the b_j remain unchanged except for b_{u+1} which decreases by 1. Hence the value of the desired sum is unchanged. By repeating this increment process (in decreasing order of the subscript so as to maintain a non-decreasing sequence) as long as necessary, we will eventually arrive at the constant sequence which gives the desired result.

Alternatively, we construct a $q \times p$ rectangle with columns 1 to q and rows 1 to p forming pq unit white squares. In the ith row, for each i, we blacken the first a_i squares. Then in the jth column the number of white squares remaining is equal to the number of the a_i strictly less than j. Thus this number is $b_j - 1$. Whence,

$$a_1 + a_2 + \cdots + a_q + (b_1 - 1) + (b_2 - 1) + \cdots + (b_p - 1) = pq$$

giving the desired result.

Note that the special case of the above result where $a_i = i$ for $1 \leqslant i \leqslant k$ gives the summation formula

$$1 + 2 + \cdots + k = \frac{k(k + 1)}{2}.$$

C. & P./10. (1984/4). A difficult mathematical competition consisted of a Part I and a Part II with a combined total of 28 problems. Each contestant solved 7 problems altogether. For each pair of problems, there were exactly two contestants who solved both of them. Prove that there was a contestant who, in Part I, solved either no problems or at least four problems.

Solution. Let r be the number of contestants who solved an arbitrary problem. These contestants also solved $6r$ other problems counting multiplicities. Since each of the remaining 27 problems is counted twice in the $6r$, we have $r = 2 \times 27/6 = 9$. Hence each problem was solved by 9 contestants and the number of contestants $= 9 \times 28/7 = 36$.

Our proof is now indirect so we assume the contrary that every contestant solved either 1, 2, or 3 problems in Part I. Now let n be the number of problems in Part I, and x, y, and z be the respective numbers of contestants who solved 1, 2, and 3 problems in Part I. Then the total number of contestants is

$$(1) \qquad\qquad x + y + z = 36.$$

Since each problem was solved by 9 contestants,

$$(2) \qquad\qquad x + 2y + 3z = 9n.$$

Assuming that $n \geqslant 2$, we claim that

$$(3) \qquad\qquad y + 3z = 2\binom{n}{2}.$$

To see why (3) holds, imagine that each contestant gets a badge for each pair of Part I problems he or she solved. Then each of the y contestants who solved two problems got one badge, and each of the z contestants who solved three problems got $\binom{3}{2} = 3$ badges. Thus the total number of badges handed out is given by the left side of (3). A different expression for the same number is obtained from the condition that each pair of problems was solved by exactly two contestants, and this is on the right side of (3).

Multiplying equations (1), (2), and (3) by -3, 3, and -2, respectively, and adding, we obtain

$$y = -2n^2 + 29n - 108 = -2(n - 29/4)^2 - 23/8 < 0.$$

Since the latter result is impossible, we obtain a contradiction. Therefore, there was a least one contestant who solved either no problems or at least 4 problems in Part I.

Comment: The conditions on the difficult mathematical competition can actually be realized from the following $(28, 36, 9, 7, 2)$ block design (see M. Hall, *Combinatorial Theory*, Blaisdell, N.Y., 1967, p. 293, #33):

```
a b d h r y z       b f j n t v w       d i l p v x *
a b i l n q u       b h l o s t x       d k n o u v z
a c d j t * #       b j k m o q *       e g i j n x y
a c f g o q x       b k p w x y #       e j o p r s u
a e h m n p t       c d n p q s w       e l m q w z #
a e o v w y *       c e f h k l v       f h p q u y *
a f i k r s w       c h j u w x z       f i o p t z #
a g j k l p z       c i k m t u y       f m n r x z *
a m s u v x #       c l n o r y #       g h k n s * #
b c e i s z *       d e k q r t x       g l r t u w *
b c g m p r v       d f j l m s y       g q s t v y z
b d e f g u #       d g h i m o w       h i j q r v #
```

Identify the 28 problems in the competition with the 28 elements of the design and the contestants with the 36 rows of the design (here printed in 3 columns of 12 rows each), with the elements in a row representing the problems solved by each contestant.

For a more general result, see H. L. Abbott, A. Liu, *On property B(4)*, Ars Combinatoria, to appear. A special case of the general result appears as Problem 1063, Crux Mathematicorum 11 (1985) p. 220; 13 (1987) pp. 20–22.

C. & P./11. (1983/1).   On a given circle, six points A, B, C, D, E, and F are chosen at random, independently and uniformly with respect to arc length. Determine the probability that the two triangles ABC and DEF are disjoint, i.e., have no common points.

Solution.   We can ignore coincidences of any of the set of points since these probabilities are zero by uniformity of the probability distribution with respect to arc length. The number of circular arrangements (with cyclic permutations factored out) is $6!/6 = 5!$, and each of these permutations has the same probability by the symmetry of the distribution. The number of different arrangements such that ABC is disjoint from DEF is $3! \times 3!$, arising from the internal ordering among A, B, C, and independently among D, E, F. Hence the probability that ABC and DEF are disjoint is $3! \times 3!/5! = 3/10$.

More generally, if we chose $m + n$ points $A_1, A_2, \ldots, A_m, B_1, B_2, \ldots, B_n$ at random, then using a similar argument, the probability that the polygons $A_1 A_2 \ldots A_m$ and $B_1 B_2 \ldots B_n$ are disjoint is $m!n!/(m + n - 1)!$.

C. & P./12. (1973/3). Three distinct vertices are chosen at random from the vertices of a given regular polygon of $(2n + 1)$ sides. If all such choices are equally likely, what is the probability that the center of the given polygon lies in the interior of the triangle determined by the three chosen random points?

Solution. Let the vertices of the polygon in order be $V_0, V_1, \ldots, V_{2n}$. We can assume that the first vertex chosen is fixed at V_0. Then the number of ways of picking two more vertices is $\binom{2n}{2}$. Now if one of the remaining two random vertices is $V_k, 1 \leqslant k \leqslant n$, there will be k triangles possible that contain the center. To see this, consider the following figure.

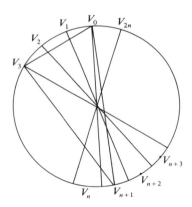

If say $k = 3$, then the only possible triangles with vertices V_0, V_3 which contain the center are $V_0 V_3 V_{n+1}$, $V_0 V_3 V_{n+2}$, and $V_0 V_3 V_{n+3}$. Thus the number of favorable cases is $\sum_{k=1}^{n} k = n(n + 1)/2$. Finally, the desired probability is

$$P = \frac{n(n + 1)}{2\binom{2n}{2}} = \frac{n + 1}{2(2n - 1)}.$$

C. & P./13. (1975/5).   A deck of n playing cards, which contains three aces, is shuffled at random (it is assumed that all possible card distributions are equally likely). The cards are then turned up one by one from the top until the second ace appears. Prove that the expected (average) number of cards to be turned up is $(n + 1)/2$.

Solution.   Let x_1, x_2, x_3 denote possible positions of the three aces in any deal. Then another equally probable deal would be the reverse of the previous deal (counting from the bottom). Hence $x_2' = n + 1 - x_2$, so that regardless of whether n is even or not, the average position is

$$\{x_2 + (n + 1 - x_2)\}/2 = (n + 1)/2.$$

It follows in a similar way that if there are r aces in a deck of n cards, and E_j denotes the expected number of cards to be turned up until the jth ace appears, then

$$E_j + E_{r+1-j} = n + 1.$$

More generally, it can be shown that $E_j = j(n + 1)/(r + 1)$ in the following manner. In a random deal, let N_i denote the number of cards between the $(i - 1)$th and the ith ace, $i = 2, 3, \ldots, r$. In addition, define N_1 to be the number of cards before the first ace and N_{r+1} is the number of cards after the last ace. The N_i are random variables, and by symmetry every $(r + 1)$-tuple $(n_1, n_2, \ldots, n_{r+1})$ of possible values of $(N_1, N_2, \ldots, N_{r+1})$ where $0 \leqslant n_k \leqslant n - r$ and $n_1 + n_2 + \cdots + n_{r+1} = n - r$, is equally likely. Consequently, the expected value $E(N_j)$ of N_j is the same for all j. Since $N_1 + N_2 + \cdots + N_{r+1} = n - r$, it follows that

$$E(N_1) + E(N_2) + \ldots E(N_{r+1}) = n - r,$$

and hence $E(N_j) = (n - r)/(r + 1)$. Then

$$E_1 = E(N_1) + 1 = (n - r)/(r + 1) + 1 = (n + 1)/(r + 1),$$
$$E_2 = E(N_1) + 1 + E(N_2) + 1 = 2(n + 1)/(r + 1), \quad \text{etc.}$$

Also, see the second solution of problem 1981/2 [39; pp. 32–33].

A continuous version of the above is given by the following:

Suppose r buses arrive randomly at a given station between times 0 and T, independently and uniformly with respect to time. What is the expected time E_j of arrival of the jth bus? As now expected, $E_j = jT/(r + 1)$. A proof by symmetry and extensions are given in M. S. Klamkin, D. J. Newman, *Inequalities and Identities for Sums and Integrals*, Amer. Math. Monthly 83 (1976), pp. 26–30.

C. & P./14. (1972/3).   A random number selector can only select one of the nine integers $1, 2, \ldots, 9$, and it makes these selections with equal probability. Determine the probability that after n selections ($n > 1$), the product of the n numbers selected will be divisible by 10.

Solution.   In order for the product to be divisible by 10, there must be at least one 5 and at least one even number among the n numbers that have been selected. Let A denote the event of obtaining at least one 5 and let B denote the event of obtaining at least one even number in the n numbers selected. Then if AB, A', $P(E)$ denote the intersection of A and B, the complement of A, and the probability of the event E, respectively, we have

$$P(AB) = 1 - P(A') - P(B') + P(A'B').$$

$$P(AB) = 1 - (8/9)^n - (5/9)^n + (4/9)^n.$$

Exercise: Redo the problem if the ten integers $0, 1, \ldots, 9$ can be selected with equal probability.

Appendix

USAMO Winners

1972

1. James Saxe, Albany High School, Albany, New York
2. Thomas Hemphill, James Monroe High School, Sepulveda, California
2. David Vanderbilt, Garden City High School, Garden City, New York
3. Paul Harrington, Paul V. Moore High School, Central Square, New York
3. Arthur Rubin, West Lafayette High School, West Lafayette, Indiana
4. David Anick, Ranney School, New Shrewsbury, New Jersey
4. Steven Raher, Central High School, Sioux City, Iowa
5. James Shearer, Livermore High School, Livermore, California

1973

1. Sheldon Katz, Brooklyn Technical High School, Brooklyn, New York
2. Eric S. Lander, Stuyvesant High School, Manhattan, New York
3. Gerhard Arenstorf, Peabody Demonstration School, Nashville, Tennessee
4. Martin D. Hirsch, Grant High School, Van Nuys, California
5. David Anick, Ranney School, New Shrewsbury, New Jersey
6. Ernest S. Davis, Classical High School, Providence, Rhode Island
6. Bruce E. Hajek, Willowbrook High School, Villa Park, Illinois
7. Karl C. Rubin, Woodrow Wilson High School, Washington, D.C.

1974

1. Paul Zeitz, Stuyvesant High School, Manhattan, New York
2. Stephen Modzelewski, Shady Side Academy, Pittsburgh, Pennsylvania
3. Gerhard Arenstorf,† Peabody Demonstration School, Nashville, Tennessee

†He suffered a fatal accident in 1974. In commemoration, an Arenstorf Medal is awarded to USAMO winners each year.

4. Thomas Nisonger, Walt Whitman High School, Bethesda, Maryland
5. Eric S. Lander, Stuyvesant High School, Manhattan, New York
6. David Barton, Berkeley High School, Berkeley, California
7. George Gilbert, Washington Lee High School, Arlington, Virginia
8. Paul Herdeg, Hamilton-Wenham High School, Hamilton, Massachusetts

1975

1. Paul Herdeg, Hamilton-Wenham High School, Hamilton, Massachusetts
2. Miller Puckette, Sewanee Academy, Sewanee, Tennessee
3. Steven Tschantz, Irvington High School, Fremont, California
4. Paul Vojta, Southwest Secondary School, Minneapolis, Minnesota
5. Russell Lyons, Lexington High School, Lexington, Massachusetts
6. Bernard Beard, Keystone School, San Antonio, Texas
7. Reed Kelly, Stuyvesant High School, Manhattan, New York
8. Alan Geller, Ridgewood High School, Ridgewood, New Jersey

1976

1. Mark Kleiman, Stuyvesant High School, Manhattan, New York
1. Adam Stephanides, University of Chicago Laboratory School, Chicago, Illinois
2. Paul Herdeg, Hamilton-Wenham High School, Hamilton, Massachusetts
3. Randall Dougherty, W. Y. Woodson High School, Fairfax, Virginia
3. Daniel Knierim, Hiram Johnson Sr. High School, Sacramento, California
4. Richard Mifflin, Stratford Sr. High School, Houston, Texas
5. Miller Puckette, Sewanee Academy, Sewanee, Tennessee
6. Reed Kelly, Stuyvesant High School, Manhattan, New York

1977

1. Mark Kleiman, Stuyvesant High School, Manhattan, New York
2. Randall Dougherty, W. Y. Woodson High School, Fairfax, Virginia
3. Peter Shor, Tamalpais High School, Hill Valley, California
4. Michael Larsen, Lexington High School, Lexington, Massachusetts
5. James Propp, North Senior High School, Great Neck, New York
6. Ronald Kaminsky, Albany High School, Albany, New York
7. Victor Milenkovic, New Trier High School East, Winnetka, Illinois
8. Paul Weiss, Stuyvesant High School, Manhattan, New York

1978

1. Randall Dougherty, W. Y. Woodson High School, Fairfax, Virginia
2. Ehud Reiter, T. S. Wooton High School, Rockville, Maryland

3. Mark Kleiman, Stuyvesant High School, Manhattan, New York
4. Michael Larsen, Lexington High School, Lexington, Massachusetts
5. Daniel Block, Bellport High School, Brookhaven, New York
6. David Montana, Lawrenceville High School, Lawrenceville, New Jersey
7. Victor Milenkovic, New Trier High School East, Winnetka, Illinois
8. Charles Walter, Centennial High School, Champaign, Illinois

1979

1. Michael Finn, Lake Braddock Secondary School, Burke, Virginia
2. Bruce Smith, Terra Linda High School, San Rafael, California
3. Michael Larsen, Lexington High School, Lexington, Massachusetts
4. Lawrence Penn, North High School, Great Neck, New York
5. Ronald Kaminsky, Albany High School, Albany, New York
6. Mark Pleszkoch, Osbourne High School, Manassas, Virginia
7. Randy Ekl, Norwin Senior High School, North Huntington, Pennsylvania
8. Richard Agin, University of Chicago Laboratory School, Chicago, Illinois

1980

1. Michael Larsen, Lexington High School, Lexington, Massachusetts
2. Eric Carlson, Munster High School, Munster, Indiana
3. Michael Finn, Lake Braddock Secondary School, Burke, Virginia
4. Bruce Smith, Terra Linda High School, San Rafael, California
5. Jeremy D. Primer, Columbia High School, Maplewood, New Jersey
6. Paul Feldman, Stuyvesant High School, Manhattan, New York
7. Daniel Scales, Westwood High School, Westwood, Massachusetts
8. David Ash, Fort William Collegiate, Thunder Bay, Ontario

1981

1. Noam Elkies, Stuyvesant High School, Manhattan, New York
1. Gregg N. Patruno, Stuyvesant High School, Manhattan, New York
2. Richard A. Stong, Albermarle High School, Charlottesville, Virginia
3. Benji N. Fisher, Bronx High School of Science, Bronx, New York
4. Jeremy D. Primer, Columbia High School, Maplewood, New Jersey
5. Brian R. Hunt, Montgomery Blair High School, Silver Spring, Maryland
6. David S. Yuen, Lane Technical High School, Chicago, Illinois
7. James R. Roche, Hill-Murray High School, St. Paul, Minnesota

1982

1. Noam Elkies, Stuyvesant High School, Manhattan, New York
2. Douglas Jungreis, George W. Hewlett High School, Hewlett, New York

3. Brian R. Hunt, Montgomery Blair High School, Silver Spring, Maryland
4. Tsz Mei Ko, Long Island City High School, Long Island City, New York
5. Washington Taylor IV, Cambridge Rindge and Latin School, Cambridge, Massachusetts
6. Vance Maverick, Harvard School, North Hollywood, California
7. David Zinke, Glencoe District High School, Clencoe, Ontario
8. Edith Star, Springfield High School, Philadelphia, Pennsylvania

1983

1. John Steinke, James Madison High School, San Antonio Texas
2. Michael Reid, Brooklyn Technical High School, Brooklyn, New York
3. James Yeh, Mountain Brook High School, Mountain Brook, Alabama
4. Jeremy Kahn, Hunter College High School, Manhattan, New York
5. Steve Newman, Roeper School, Bloomfield Hills, Michigan
6. Douglas Jungreis, George W. Hewlett High School, Hewlett, New York
7. Douglas Davidson, Langley High School, McLean, Virginia
8. John O'Neil, Lawrenceville School, Lawrenceville, New Jersey

1984

1. David Grabiner, Claremont High School, Claremont, California
2. Douglas Davidson, Langley High School, McLean, Virginia
3. David Moews, Edwin O. Smith High School, Storrs, Connecticut
4. Joseph Keane, Fox Chapel High School, Pittsburgh, Pennsylvania
5. Steve Newman, Roeper School, Bloomfield Hills, Michigan
6. Michael Reid, Brooklyn Technical High School, Brooklyn, New York
7. Andrew Chinn, S. F. Austin High School, Austin, Texas
8. William Jockusch, University High School, Urbana, Illinois

1985

1. Joseph Keane, Fox Chapel High School, Pittsburgh, Pennsylvania
2. Waldemar Horwart, Hoffman Estates High School, Hoffman Estates, Illinois
2. John Overdeck, Wilde Lake High School, Columbia, Maryland
3. Yeh-Ching Tung, Saratoga High School, Saratoga, California
4. Bjorn Poonen, Winchester High School, Winchester, Massachusetts
5. Zinkoo Han, South Shore High School, Brooklyn, New York
6. Jeremy Kahn, Hunter College High School, Manhattan, New York
7. John Dalbeck, Austintown Fitch High School, Youngstown, Ohio

1986

1. Joseph Keane, Fox Chapel High School, Pittsburgh, Pennsylvania
2. David Grabiner, Claremont High School, Claremont, California

3. Ravi Vakil, Martingrove Collegiate Institute, Islington, Ontario
4. Darian Lefkowitz, Stuyvesant High School, Manhattan, New York
5. Will Croff, Kalamazoo Central High School, Kalamazoo, Michigan
6. John Bulten, Holland High School, Tulsa, Oklahoma
6. Eric Wepsic, Boston Latin School, Boston, Massachusetts
7. Jeremy Kahn, Hunter College High School, Manhattan, New York

For brief accounts of a follow up of the academic progress and careers of the winners, see Nura D. Turner, *Who's Who of USA Olympiad Participants 1972–1986*, Fort Orange Press, Albany, N.Y., and the USAMO Newsletter 1 (1988) pp. 1–49 (available for a small fee from Nura D. Turner, Dept. of Math. and Stat., SUNY at Albany, Albany, N.Y. 12222).

List of Symbols

$[ABC]$	area of $\triangle ABC$
$\simeq$ or $\approx$	approximately equal to
$\cong$	congruent (in geometry)
$a \equiv b \pmod{p}$	$a - b$ is divisible by p; *Congruence*, see Glossary
$a \not\equiv b \pmod{p}$	$a - b$ is not divisible by p
$\equiv$	identically equal to
$[x]$ or $\lfloor x \rfloor$	integer part of x, i.e. greatest integer not exceeding x
$\lceil x \rceil$	least integer greater than or equal to x
$\binom{n}{k}$, $C(n, k)$	binomial coefficient, see Glossary; also the number of combinations of n things, k at a time
(n, k), G.C.D. of n, k	the greatest common divisor of n and k
$p \mid n$	p divides n
$p \nmid n$	p does not divide n
$n!$	n factorial $= 1 \cdot 2 \cdot 3 \cdot \ldots (n-1)n$, $0! = 1$
$\displaystyle\prod_{i=1}^{n} a_i$	the product $a_1 \cdot a_2 \cdot \cdots \cdot a_n$
$\sim$	similar in geometry
$\displaystyle\sum_{i=1}^{n} a_i$	the sum $a_1 + a_2 + \cdots + a_n$
$\circ$	$f \circ g(x) = f[g(x)]$, see *Composition* in Glossary
$K_1 \cup K_2$	union of sets K_1, K_2
$K_1 \cap K_2$	intersection of sets K_1, K_2
A. M.	arithmetic mean, see *Mean* in Glossary
G. M.	geometric mean, see *Mean* in Glossary
H. M.	harmonic mean, see *Mean* in Glossary
$[a, b]$	closed interval, i.e. all x such that $a \leqslant x \leqslant b$
(a, b)	open interval, i.e. all x such that $a < x < b$

Glossary of some frequently used terms and theorems.

Arithmetic mean (average). see *Mean*

Arithmetic mean-geometric mean inequality (A.M.-G.M. inequality).
  If $a_1, a_2, \ldots, a_n$ are n non-negative numbers, then

$$\frac{1}{n} \sum_{i=1}^{n} a_i \geq \left[\prod_{i=1}^{n} a_i \right]^{1/n} \quad \text{with equality if and only if} \quad a_1 = a_2 = \cdots = a_n.$$

Weighted arithmetic mean-geometric mean inequality.
  If, in addition, $w_1, w_2, \ldots, w_n$ are non-negative numbers (weights) whose sum is 1, then

$$\sum_{i=1}^{n} w_i a_i \geq \prod_{i=1}^{n} a_i^{w_i} \quad \text{with equality if and only if} \quad a_1 = a_2 = \cdots = a_n.$$

  For a proof, use Jensen's inequality below, applied to $f(x) = -\log x$.

Arithmetic Series. see *Series*

Binomial coefficient.

$$\binom{n}{k} = \frac{n!}{k!(n-k)!} = \binom{n}{n-k} = \text{coefficient of } y^k \text{ in the expansion } (1+y)^n.$$

Also $\binom{n+1}{k+1} = \binom{n}{k+1} + \binom{n}{k}$. (See *Binomial theorem* and List of Symbols.)

Binomial theorem.

$$(x+y)^n = \sum_{k=0}^{n} \binom{n}{k} x^{n-k} y^k, \quad \text{where}$$

$$\binom{n}{k} = \frac{n(n-1)\cdots(n-k+1)}{1 \cdot 2 \cdot \ldots \cdot (k-1)k} = \frac{n!}{k!(n-k)!}.$$

Cauchy's inequality.

For vectors $\mathbf{x}, \mathbf{y}$, $|\mathbf{x} \cdot \mathbf{y}| \leqslant |\mathbf{x}||\mathbf{y}|$; componentwise, for real numbers x_i, y_i, $i = 1, 2, \ldots, n$,

$$|x_1 y_1 + x_2 y_2 + \cdots + x_n y_n| \leqslant \left[\sum_{i=1}^{n} x_i^2 \right]^{1/2} \left[\sum_{i=1}^{n} y_i^2 \right]^{1/2}.$$

There is equality if and only if $\mathbf{x}, \mathbf{y}$ are collinear, i.e., if and only if $x_i = k y_i$, $i = 1, 2, \ldots, n$. A proof for vectors follows from the definition of dot product $\mathbf{x} \cdot \mathbf{y} = |\mathbf{x}||\mathbf{y}|\cos(\mathbf{x}, \mathbf{y})$ or by considering the discriminant of the quadratic function $q(t) = \Sigma(y_i t - x_i)^2$.

Centroid of a triangle.

Point of intersection of the medians

Ceva's theorem.

If AD, BE, CF are concurrent cevians of a triangle ABC, then

(i) $$BD \cdot CE \cdot AF = DC \cdot EA \cdot FB.$$

Conversely, if AD, BE, CF are three cevians of a triangle ABC such that (i) holds, then the three cevians are concurrent. (A *cevian* is a segment joining a vertex of a triangle with a point on the opposite side.)

Chinese remainder theorem.

Let $m_1, m_2, \ldots, m_n$ denote n positive integers that are relatively prime in pairs, and let $a_1, a_2, \ldots, a_n$ denote any n integers. Then the congruences $x \equiv a_i (\mathrm{mod}\, m_i)$, $i = 1, 2, \ldots, n$ have common solutions; any two solutions are congruent modulo $m_1 m_2 \ldots m_n$. For a proof, see [112, p. 31].

Circumcenter of $\triangle ABC$.

Center of circumscribed circle of $\triangle ABC$

Circumcircle of $\triangle ABC$.

Circumscribed circle of $\triangle ABC$

Complex numbers.

Numbers of the form $x + iy$, where x, y are real and $i = \sqrt{-1}$.

Composition of functions.

$F(x) = f \circ g(x) = f[g(x)]$ is the composite of functions, f, g, where the range of g is the domain of f.

Congruence.

$a \equiv b (\mathrm{mod}\, p)$ read "*a is congruent to b modulo p*" means that $a - b$ is divisible by p.

Concave function.

$f(x)$ is concave if $-f(x)$ is convex; see *Convex function*.

Convex function.
　　A function $f(x)$ is convex in an interval I if for all x_1, x_2 in I and for all non-negative weights w_1, w_2 with sum 1,

$$w_1 f(x_1) + w_2 f(x_2) \geq f(w_1 x_1 + w_2 x_2).$$

　　Geometrically this means that the graph of f between $(x_1, f(x_1))$ and $(x_2, f(x_2))$ lies below its secants.
　　We state the following useful facts:
　　1. A continuous function which satisfies the above inequality for $w_1 = w_2 = 1/2$ is convex.
　　2. A twice differentiable function f is convex if and only if $f''(x)$ is non-negative in the interval in question.
　　3. The graph of a differentiable convex function lies above its tangents.
　　For an even more useful fact, see *Jensen's inequality*.

Convex hull of a pointset S.
　　The intersection of all convex sets containing S

Convex pointset.
　　A pointset S is convex if, for every pair of points P, Q in S, all points of the segment PQ are in S.

Cross product (vector product) $\mathbf{x} \times \mathbf{y}$ *of two vectors.* see *Vectors.*

Cyclic polygon.
　　Polygon that can be inscribed in a circle.

de Moivre's theorem.
　　$(\cos \theta + i \sin \theta)^n = \cos n\theta + i \sin n\theta$. For a proof, see NML vol. 27, p. 49.

Determinant of a square matrix M (det M).
　　A multi-linear function $f(C_1, C_2, \ldots, C_n)$ of the columns of M with the properties

$$f(C_1, C_2, \ldots, C_i, \ldots, C_j, \ldots, C_n) = -f(C_1, C_2, \ldots, C_j, \ldots, C_i, \ldots, C_n)$$

and det $I = 1$. Geometrically, $\det(C_1, C_2, \ldots, C_n)$ is the signed volume of the n-dimensional oriented parallelepiped with coterminal side vectors $C_1, C_2, \ldots C_n$.

Dirichlet's principle. see *Pigeonhole principle.*

Dot product (scalar product) $\mathbf{x} \cdot \mathbf{y}$ *of two vectors.* see *Vectors.*

Escribed circle. see *Excircle.*

Euclid's algorithm.
　　A process of repeated divisions yielding the greatest common divisor of two integers, $m > n$:

$$m = nq_1 + r_1, \quad q_1 = r_1 q_2 + r_2, \quad \ldots, \quad q_k = r_k q_{k+1} + r_{k+1};$$

the last non-zero remainder is the GCD of m and n. For a detailed discussion, see e.g. C. D. Olds, *Continued Fractions*, NML vol. 9 (1963), p. 16.

Euler's extension of Fermat's theorem. see *Fermat's theorem.*

Euler's theorem on the distance d between in- and circumcenters of a triangle.
$d = \sqrt{R^2 - 2rR}$, where r, R are the radii of the inscribed and circumscribed circles.

Excircle of $\triangle ABC$.
A circle that touches one side of the triangle internally and the other two (extended) externally.

Fermat's Theorem.
If p is a prime, $a^p \equiv a \pmod{p}$.
Euler's extension of $-$.
If m is relatively prime to n, then $m^{\phi(n)} \equiv 1 \pmod{n}$, where the Euler $\phi(n)$ function is defined to be the number of positive integers $\leq n$ and relatively prime to n. There is a simple formula for ϕ:

$$\phi(n) = n \prod \left(1 - \frac{1}{p_j} \right), \quad \text{where } p_j \text{ are distinct prime factors of } n.$$

Fundamental summation formula.
Our name for the telescoping sum formula, to point out its similarity to the Fundamental Theorem of Calculus. see *Summation of Series.*

Geometric mean. see *Mean, geometric.*

Geometric series. see *Series, geometric.*

Harmonic mean. see *Mean, harmonic.*

Heron's formula.
The area of $\triangle ABC$ with sides a, b, c is
$[ABC] = \sqrt{s(s-a)(s-b)(s-c)}$, where $s = \frac{1}{2}(a + b + c)$.

Hölder's inequality.
If a_i, b_i are non-negative numbers, and if p, q are positive numbers such that $(1/p) + (1/q) = 1$, then
$$a_1 b_1 + a_2 b_2 + \cdots + a_n b_n$$

$$\leq (a_1^p + a_2^p + \cdots + a_n^p)^{1/p} (b_1^q + b_2^q + \cdots + b_n^q)^{1/q}$$

with equality if and only if $a_i = kb_i$, $i = 1, 2, \ldots, n$. Cauchy's inequality corresponds to the special case $p = q = 2$.

Homogeneous.

$f(x, y, z, \ldots)$ is homogeneous of degree k if

$$f(tx, ty, tz, \ldots) = t^k f(x, y, z, \ldots).$$

A system of linear equations is called homogeneous if each equation is of the form $f(x, y, z, \ldots) = 0$ with f homogeneous of degree 1.

Homothety.

A dilatation (simple stretch or compression) of the plane (or space) which multiplies all distances from a fixed point, called the *center of homothety* (or *similitude*), by the same factor $\lambda \neq 0$. This mapping (transformation) is a similarity which transforms each line into a parallel line, and the only point unchanged (invariant) is the center. Conversely, if any two similar figures have their corresponding sides parallel, then there is a homothety which transforms one of them into the other, and the center of homothety is the point of concurrence of all lines joining pairs of corresponding points. Two physical examples are a photo enlarger and a pantograph.

Incenter of $\triangle ABC$.

Center of inscribed circle of $\triangle ABC$

Incircle of $\triangle ABC$.

Inscribed circle of $\triangle ABC$

Inequalities.

A.M. – G.M.—see *Arithmetic mean*
A.M. – H.M.—see *Mean, Harmonic*
Cauchy—see *Cauchy's-*
H.M. – G.M.—see *Mean, Harmonic*
Hölder—see *Hölder's-*
Jensen—see *Jensen's-*
Power mean—see *Power mean-*
Schur—see *Schur's-*
Triangle—see *Triangle-*

Inverse function.

$f: X \to Y$ has an inverse f^{-1} if for every y in the range of f there is a unique x in the domain of f such that $f(x) = y$; then $f^{-1}(y) = x$, and $f^{-1} \circ f$, $f \circ f^{-1}$ are the identity functions. See also *Composition.*

Irreducible polynomial.

A polynomial $g(x)$, not identically zero, is irreducible over a field F if there is no factoring, $g(x) = r(x)s(x)$, of $g(x)$ into two polynomials $r(x)$ and $s(x)$ of positive degrees over F. For example, $x^2 + 1$ is irreducible over the real number field, but reducible, $(x + i)(x - i)$, over the complex number field.

Isoperimetric theorem for polygons.

Among all n-gons with given area, the regular n-gon has the smallest perimeter. Dually, among all n-gons with given perimeter, the regular n-gon has the largest area.

Jensen's inequality.

If $f(x)$ is convex in an interval I and $w_1, w_2, \ldots, w_n$ are arbitrary non-negative weights whose sum is 1, then

$$w_1 f(x_1) + w_2 f(x_2) + \cdots + w_n f(x_n)$$
$$\geq f(w_1 x_1 + w_2 x_2 + \cdots + w_n x_n)$$

for all x_i in I.

Matrix.

A rectangular array of number (a_{ij})

Mean of n numbers.

Arithmetic mean (*average*) $= A.M. = \dfrac{1}{n} \sum\limits_{i=1}^{n} a_i$

Geometric mean $= G.M. = \sqrt[n]{a_1 a_2 \cdots a_n}$, $a_i \geq 0$

Harmonic mean $= H.M. = \left(\dfrac{1}{n} \sum\limits_{i=1}^{n} \dfrac{1}{a_i} \right)^{-1}$, $a_i > 0$

A.M. – G.M. – H.M. inequalities

$A.M. \geq G.M. \geq H.M.$ with equality if and only if all n numbers are equal.

Power mean $= P(r) = \left[\dfrac{1}{n} \sum\limits_{i=1}^{n} a_i^r \right]^{1/r}$, $a_i > 0$, $r \neq 0$, $|r| < \infty$

$$= [\Pi a_i]^{1/n} \quad \text{if } r = 0$$
$$= \min(a_i) \quad \text{if } r = -\infty$$
$$= \max(a_i) \quad \text{if } r = \infty$$

Special cases: $P(0) = G.M.$, $P(-1) = H.M.$, $P(1) = A.M.$

It can be shown that $P(r)$ is continuous on $-\infty \leq r \leq \infty$, that is

$$\lim_{r \to 0} P(r) = [\Pi a_i]^{1/n}, \quad \lim_{r \to -\infty} P(r) = \min(a_i),$$
$$\lim_{r \to \infty} P(r) = \max(a_i).$$

– inequality.

$P(r) \leq P(s)$ for $-\infty \leq r < s \leq \infty$, with equality if and only if all the a_i are equal. For a proof, see [120, pp. 76–77.]

Menelaus' theorem.

If D, E, F, respectively, are three collinear points on the sides BC, CA, AB of a triangle ABC, then

(i) $\qquad\qquad BD \cdot CE \cdot AF = -DC \cdot EA \cdot FB.$

Conversely, if D, E, F, respectively are three points on the sides BC, CA, AB of a triangle ABC suitably extended, such that (i) holds, then D, E, F are collinear.

Orthocenter of $\triangle ABC$.
Point of intersection of altitudes of $\triangle ABC$

Periodic function.
$f(x)$ is periodic with period a if $f(x + a) = f(x)$ for all x.

Pigeonhole principle (Dirichlet's box principle).
If n objects are distributed among $k < n$ boxes, some box contains at least two objects.

Polynomial in x of degree n.
Function of the form $P(x) = \sum_{i=0}^{n} c_i x^i$, $c_n \neq 0$.
Irreducible-see *Irreducible-*

Radical axis of two non-concentric circles.
Locus of points of equal powers with respect to the two circles. (If the circles intersect, it is the line containing the common chord.)

Radical center of three circles with non-collinear centers.
Common intersection of the three radical axes of each pair of circles.

Root of an equation.
Solution of an equation

Roots of unity.
Solutions of the equation $x^n - 1 = 0$.

Schur's inequality.
$x^n(x - y)(x - z) + y^n(y - z)(y - x) + z^n(z - x)(z - y) \geq 0$, for all real $x, y, z, n \geq 0$. See [120].

Series.
Arithmetic: $\sum_{j=1}^{n} a_j$ with $a_{j+1} = a_j + d$, d the common difference.
Geometric: $\sum_{j=0}^{n-1} a_j$ with $a_{j+1} = ra_j$, r the common ratio.
Summation of—
Linearity: $\sum_{k} [aF(k) + bG(k)] = a\sum_{k} F(k) + b\sum_{k} G(k)$.
Fundamental theorem of- or *Telescoping sums theorem*;
$$\sum_{k=1}^{n} [F(k) - F(k - 1)] = F(n) - F(0)$$
(so named in analogy with Fundamental Theorem of Calculus).

By choosing F appropriately, we can obtain the following sums:

$$\sum_{k=1}^{n} 1 = n, \quad \sum_{k=1}^{n} k = \frac{1}{2}n(n+1), \quad \sum_{k=1}^{n} k^2 = \frac{1}{6}n(n+1)(2n+1),$$

$$\sum_{k=1}^{n} [k(k+1)]^{-1} = 1 - \frac{1}{n+1},$$

$$\sum_{k=1}^{n} [k(k+1)(k+2)]^{-1} = \frac{1}{4} - \frac{1}{2(n+1)(n+2)}.$$

$$\sum_{k=1}^{n} ar^{k-1} = a(1 - r^n)/(1 - r), \quad \text{the sum of a geometric series,}$$

see above.

$$\sum_{k=1}^{n} \cos 2kx = \frac{\sin nx \cos(n+1)x}{\sin x},$$

$$\sum_{k=1}^{n} \sin 2kx = \frac{\sin nx \sin(n+1)x}{\sin x}$$

Subadditive.

A function $f(x)$ is subadditive if $f(x + y) \le f(x) + f(y)$

Superadditive.

A function $g(x)$ is superadditive if $g(x + y) \ge g(x) + g(y)$

Telescoping sum. see *Series, Fundamental theorem of summation of—*

Triangles (plane)

Law of cosines: $a^2 = b^2 + c^2 - 2bc \cos A$, etc.

Law of sines: $\dfrac{a}{\sin A} = \dfrac{b}{\sin B} = \dfrac{c}{\sin C} = 2R$, ($R$ is circumradius)

Triangles (spherical)

Law of cosines: $\cos a = \cos b \cos c + \sin b \sin c \cos A$

$$\cos A = -\cos B \cos C + \sin B \sin C \cos a, \text{ etc.}$$

Law of sines:

$$\frac{\sin A}{\sin a} = \frac{\sin B}{\sin b} = \frac{\sin C}{\sin c} = \frac{2n}{\sin a \sin b \sin c} = \frac{\sin A \sin B \sin C}{2N} = \frac{N}{n}$$

where

$$2n = \sqrt{1 - \cos^2 a - \cos^2 b - \cos^2 c + 2 \cos a \cos b \cos c},$$

$$2N = \sqrt{-\cos S \cos(S - A) \cos(S - B) \cos(S - C)}, \quad (2S = A + B + C).$$

Also,

$$n = \frac{2N^2}{\sin A \sin B \sin C}, \qquad N = \frac{2n^2}{\sin a \sin b \sin c}.$$

Trigonometric identities.

$$\left.\begin{array}{l} \sin(x \pm y) = \sin x \cos y \pm \sin y \cos x \\ \cos(x \pm y) = \cos x \cos y \mp \sin x \sin y \end{array}\right\} \quad \text{addition formulas}$$

$$\left.\begin{array}{l} \sin nx = \cos^n x \left\{ \binom{n}{1}\tan x - \binom{n}{3}\tan^3 x + \cdots \right\} \\ \cos nx = \cos^n x \left\{ 1 - \binom{n}{2}\tan^2 x + \binom{n}{4}\tan^4 x - \cdots \right\} \end{array}\right\} \quad \begin{array}{l} \text{consequences of} \\ \text{de Moivre's} \\ \text{theorem} \end{array}$$

$$\sin 2x + \sin 2y + \sin 2z - \sin 2(x + y + z)$$
$$= 4 \sin(y + z)\sin(z + x)\sin(x + y),$$

$$\cos 2x + \cos 2y + \cos 2z + \cos 2(x + y + z)$$
$$= 4 \cos(y + z)\cos(z + x)\cos(x + y),$$

$$\sin(x + y + z)$$
$$= \cos x \cos y \cos z(\tan x + \tan y + \tan z - \tan x \tan y \tan z),$$

$$\cos(x + y + z)$$
$$= \cos x \cos y \cos z(1 - \tan y \tan z - \tan z \tan x - \tan x \tan y).$$

Vectors

One may consider an n-dimensional vector as an ordered n-tuple of real numbers: $\mathbf{x} = (x_1, x_2, \ldots, x_n)$. Its product with any real number a is the vector $a\mathbf{x} = (ax_1, ax_2, \ldots, ax_n)$. The *sum* of two vectors $\mathbf{x}$ and $\mathbf{y}$ is the vector $\mathbf{x} + \mathbf{y} = (x_1 + y_1, x_2 + y_2, \ldots, x_n + y_n)$ (parallelogram or triangle law of addition). The *dot-* or *scalar product* $\mathbf{x} \cdot \mathbf{y}$ is defined geometrically as $|\mathbf{x}||\mathbf{y}|\cos\theta$, where $|\mathbf{x}|$ denotes the length of $\mathbf{x}$, etc., and θ is the angle between the two vectors. Algebraically, the dot product is defined as the number

$$\mathbf{x} \cdot \mathbf{y} = x_1 y_1 + x_2 y_2 + \cdots + x_n y_n, \quad \text{and}$$

$$|\mathbf{x}|^2 = \mathbf{x} \cdot \mathbf{x} = x_1^2 + x_2^2 + \cdots + x_n^2.$$

In 3-dimensional space E^3, the *vector-* or *cross product* $\mathbf{x} \times \mathbf{y}$ is defined geometrically as a vector orthogonal to both $\mathbf{x}$ and $\mathbf{y}$, whose magnitude is $|\mathbf{x}||\mathbf{y}|\sin\theta$ and directed according to the right hand screw convention. Algebraically, the cross product of $\mathbf{x} = (x_1, x_2, x_3)$ and $\mathbf{y} = (y_1, y_2, y_3)$ is given by the vector

$$\mathbf{x} \times \mathbf{y} = (x_2 y_3 - x_3 y_2, x_3 y_1 - x_1 y_3, x_1 y_2 - x_2 y_1).$$

It follows from the geometric definition that the *triple scalar product* $\mathbf{x} \cdot \mathbf{y} \times \mathbf{z}$ is the signed volume of the oriented parallelepiped having $\mathbf{x}$, $\mathbf{y}$ and $\mathbf{z}$ as coterminal sides. It can conveniently be written as

$$\mathbf{x} \cdot \mathbf{y} \times \mathbf{z} = \begin{vmatrix} x_1 & x_2 & x_3 \\ y_1 & y_2 & y_3 \\ z_1 & z_2 & z_3 \end{vmatrix} = \det(\mathbf{x}, \mathbf{y}, \mathbf{z}), \text{ see also } Determinant.$$

Zero of a function $f(x)$.
Any point x for which $f(x) = 0$.

References

General:

1. B. Averbach and O. Chein, *Mathematics: Problem Solving through Recreational Mathematics*, Freeman, San Francisco, 1980.
2. W. W. R. Ball and H. S. M. Coxeter, *Mathematical Recreations*, Macmillian, N.Y., 1939.
3. A. Beck, M. N. Bleicher, and D. W. Crowe, *Excursions into Mathematics*, Worth, N.Y., 1969.
4. D. M. Campbell, *The Whole Craft of Number*, Prindle, Weber, and Schmidt, Boston, 1976.
5. R. Courant and H. Robbins, *What is Mathematics?*, Oxford University Press, Oxford, 1941.
6. S. Gudder, *A Mathematical Journey*, McGraw-Hill, N.Y., 1976.
7. D. H. Hilbert and S. Cohn-Vossen, *Geometry and the Imagination*, Chelsea, N.Y., 1952.
8. R. Honsberger, *Mathematical Gems*, The Dolciani Mathematical Expositions, Vols. I, II, IX, M.A.A., Wash., D.C., 1973, 1976, and 1985.
9. R. Honsberger, *Mathematical Morsels*, The Dolciani Mathematical Expositions, Vol. III, M.A.A., Wash., D.C., 1978.
10. R. Honsberger (ed.), *Mathematical Plums*, The Dolciani Mathematical Expositions, Vol. IV, M.A.A.,Wash., D.C., 1979.
11. Z. A. Melzac, *Companion to Concrete Mathematics*, Vols I, II, Wiley, N.Y., 1973, 1976.
12. G. Pólya, *How to Solve it*, Doubleday, N.Y., 1957.
13. G. Pólya, *Mathematical Discovery*, Vols I, II, Wiley, N.Y., 1962, 1965.
14. G. Pólya, *Mathematics and Plausible Reasoning*, Vols. I, II, Princeton University Press, Princeton, 1954.

15. J. Pottage, *Geometrical Investigations*, Addison-Wesley, Reading, 1983.

16. H. Rademacher and O. Toeplitz, *The Enjoyment of Mathematics*, Princeton University Press, Princeton, 1957.

17. A. W. Roberts and D. L. Varberg, *Faces of Mathematics*, Crowell, N.Y., 1978.

18. I. J. Schoenberg, *Mathematical Time Exposures*, M.A.A., Wash., D.C., 1983.

19. S. K. Stein, Mathematics, *The Man-Made Universe*, Freeman, San Francisco, 1976.

20. H. Steinhaus, *Mathematical Snapshots*, Oxford University Press, N.Y., 1969.

Problems:

21. G. L. Alexanderson, L. F. Klosinski, L. G. Larson, *The William Lowell Putnam Competition, Problems and Solutions*: 1965–1984, M.A.A., Wash., D.C., 1985.

22. M. N. Aref and W. Wernick, *Problems and Solutions in Euclidean Geometry*, Dover, N.Y., 1968.

23. E. Barbeau, M. S. Klamkin, and W. O. Moser, *1001 Problems in High School Mathematics*, Vols. I–V, Canadian Mathematical Society, Ottawa 1976–1985.

24. S. J. Bryant, G. E. Graham, and K. G. Wiley, *Nonroutine Problems in Algebra, Geometry, and Trigonometry*, McGraw-Hill, N.Y., 1865.

25. M. Charosh, *Mathematical Challenges*, N.C.T.M., Wash., D.C., 1965.

26. *Contest Problem Books* I, II, III, IV, Annual High School Mathematics Examinations 1950–60, 1961–65, 1966–72, 1973–82 (NML vols. 5, 17, 25, 29), M.A.A., Wash., D.C., 1961, 1966, 1973, 1983.

27. H. Dorrie, *100 Great Problems in Elementary Mathematics*, Dover, N.Y. 1965.

28. E. B. Dynkin et al, *Mathematical Problems: An Anthology*, Gordon and Breach, N.Y., 1969.

29. E. B. Dynkin, V. A. Uspenskii, *Problems in the Theory of Numbers*, D. C. Health, Boston, 1963.

30. E. B. Dynkin, V. A. Uspenskii, *Multicolor Problems*, D. C. Heath, Boston, 1963.

31. E. B. Dynkin, V. A. Uspenskii, *Random Walks*, D. C. Heath, Boston, 1963.

32. D. K. Fadeev and I. S. Sominski, *Problems in Higher Algebra*, Freeman, San Francisco, 1965.

33. A. M. Gleason, R. E. Greenwood, and L. M. Kelly, *The William Lowell Putnam Competition, Problems and Solutions*: 1938–1964. M.A.A., Wash., D. C., 1980.

34. K. Hardy and K. S. Williams, *The Green Book, 100 Practice Problems for Undergraduate Mathematics Competitions*, Integer Press, Ottawa, 1985.
35. T. J. Hill, *Mathematical Challenges II Plus Six*, N.C.T.M., Wash., D. C., 1974.
36. A. P. Hillman and G. L. Alexanderson, *Algebra through Problem Solving*, Allyn and Bacon, Boston, 1966.
37. *Hungarian Problem Books* I and II (Eötvös Competitions 1894–1928), translated by E. Rapaport, NML. Vols. 11 and 12, M.A.A., Wash., D.C., 1963.
38. *International Mathematical Olympiads, 1959–1977*, compiled and with solutions by S. L. Greitzer, NML vol. 27, M.A.A., Wash., D.C., 1978.
39. *International Mathematical Olympiads, 1978–1985 and forty supplementary problems* by M. S. Klamkin, NML vol. 31, M.A.A., Wash., D.C.
40. G. Klambauer, *Problems and Propositions in Analysis*, Dekker, N.Y., 1979.
41. L.C. Larson, *Problem-Solving Through Problems*, Springer-Verlag, N.Y., 1983.
42. M. Lehtinen (ed.), *26th International Mathematical Olympiad, Results and Problems*, Helsinki, 1985.
43. V. Lidsky et al., *Problems in Elementary Mathematics*, Mir, Moscow, 1973.
44. L. Lovász, *Combinatorial Problems and Exercises*, North-Holland, Amsterdam, 1979.
45. F. Mosteller, *Fifty Challenging Problems in Probability with Solutions*, Addison-Wesley, Reading, 1965.
46. D. J. Newman, *A Problem Seminar*, Springer-Verlag, N.Y., 1982.
47. G. Pólya and J. Kilpatrick, *The Stanford Mathematics Problem Book*, Teachers College Press, N.Y., 1974.
48. G. Pólya and G. Szegö, *Problems and Theorems in Analysis*, Vols. I, II, Springer-Verlag, N.Y., 1976.
49. D. O. Shklarsky, N. N. Chentzov, and I. M. Yaglom, *Selected Problems and Theorems in Elementary Mathematics*, Mir, Moscow, 1979.
50. W. Sierpinski, *A Selection of Problems in the Theory of Numbers*, Pergamon, Oxford, 1964.
51. W. Sierpinski, *250 Problems in Elementary Number Theory*, American Elsevier, N.Y., 1970.
52. H. Steinhaus, *One Hundred Problems in Elementary Mathematics*, Basic Books, N.Y., 1964.
53. S. Straszewicz, *Mathematical Problems and Puzzles from the Polish Mathematical Olympiads*, Pergamon, Oxford, 1965.

54. G. Szász et al (eds.), *Contests in Higher Mathematics*, *Hungary* 1949–1961, Akademiai Kiado, Budapest, 1968.
55. H. Tietze, *Famous Problems of Mathematics*, Greylock Press, N.Y., 1965.
56. I. Tomescu, *Problems in Combinatorics and Graph Theory*, Wiley, N.Y., 1985.
57. N. Y. Vilenkin, *Combinatorics*, Academic Press, N.Y., 1971.
58. I. M. Yaglom and V. G. Boltyanskii, *Convex Figures*, Holt, Rinehart and Winston, N.Y., 1961.
59. A. M. Yaglom and I. M. Yaglom, *Challenging Mathematical Problems with Elementary Solutions*, Vols. I, II, Holden-Day, San Francisco, 1964, 1967.

Algebra:

60. S. Barnard and J. M. Child, *Higher Algebra*, Macmillan, London, 1939.
61. G. Chrystal, *Algebra*, Vol. I, II, Chelsea, N.Y., 1952.
62. C. V. Durell and A. Robson, *Advanced Algebra*, Vols. I, II, III, Bell, London, 1964.
63. H. S. Hall and S. R. Knight, *Higher Algebra*, Macmillan, London, 1932.
64. A. Mostowski and M. Stark, *Introduction to Higher Algebra*, Pergamon, Oxford, 1964.
65. J. V. Uspensky, *Theory of Equations*, McGraw-Hill, N.Y., 1945.

Combinatorics and Probabilty:

66. C. Berge, *Principles of Combinatorics*, Academic Press, N.Y., 1971.
67. R. A. Brualdi, *Introductory Combinatorics*, Elsevier North-Holland, N.Y., 1977.
68. D. I. A. Cohen, *Basic Techniques of Combinatorial Theory*, Wiley, N.Y., 1978.
69. L. Comptet, *Advanced Combinatorics*, Reidel, Dordrecht, 1974.
70. J. W. Moon, *Topics on Tournaments*, Holt, Rinehart and Winston, N.Y., 1968.
71. J. Riordan, *Introduction to Combinatorial Analysis*, Wiley, N.Y., 1958.
72. I. Tomescu, *Introduction to Combinatorics*, Collet's, Romania, 1975.
73. A. Tucker, *Applied Combinatorics*, Wiley, N.Y., 1980.
74. E. Borel, *Elements of the Theory of Probability*, Prentice-Hall, N.J., 1950.
75. W. Burnside, *Theory of Probability*, Dover, N.Y., 1959.

76. M. M. Eisen and C. A. Eisen, *Probability and its Applications*, Quantum, N.Y., 1975.
77. W. Feller, *An Introduction to Probability Theory and its Applications*, Vol. I, Wiley, N.Y., 1961.
78. M. A. Goldberg, *An Introduction to Probability Theory with Statistical Applications*, Plenum, N.Y., 1984.
79. S. Goldberg, *Probability, An Introduction*, Prentice-Hall, N.J., 1960.
80. F. Mosteller, R. E. K. Rourke, and G. B. Thomas, Jr., *Probability and Statistics*, Addison-Wesley, Reading, 1961.
81. J. V. Uspensky, *Introduction to Mathematical Probability*, McGraw-Hill, N.Y., 1937.
82. W. A. Whitworth, *Choice and Chance*, Hafner, N.Y., 1948.

Geometry (Plane):
83. P. Andreev and E. Shuvalova, *Geometry*, Mir, Moscow, 1974.
84. N. Altshiller-Court, *College Geometry*, Barnes and Noble, N.Y., 1952.
85. H. S. M. Coxeter, *Introduction to Geometry*, Wiley, N.Y., 1969.
86. D. R. David, *Modern College Geometry*, Addison-Wesley, Reading, 1949.
87. C. W. Dodge, *Euclidean Geometry and Transformations*, Addison-Wesley, Reading, 1972.
88. C. V. Durell, *Modern Geometry*, Macmillan, London, 1952.
89. C. V. Durell, *Projective Geometry*, Macmillan, London, 1952.
90. H. Eves, *A Survey of Geometry*, Vols. I, II, Allyn and Bacon, Boston, 1963.
91. H. G. Forder, *Higher Course Geometry*, Cambridge University Press, Cambridge, 1931.
92. L. I. Golovina and I. M. Yaglom, *Induction in Geometry*, Mir, Moscow, 1979.
93. H. W. Guggenheimer, *Plane Geometry and Its Groups*, Holden-Day, San Francisco, 1967.
94. H. Hadwiger, H. Debrunner, V. Klee, *Combinatorial Geometry in the Plane*, Holt, Rinehart and Winston, N.Y., 1964.
95. R. A. Johnson, *Advanced Euclidean Geometry*, Dover, N.Y., 1960.
96. D. C. Kay, *College Geometry*, Holt, Rinehart & Winston, N.Y., 1969.
97. Y. I. Lyubich, L. A. Shor, *The Kinematic Method in Geometrical Problems*, Mir, Moscow, 1980.
98. Z. A. Melzak, *Invitation to Geometry*, Wiley, N.Y., 1983.
99. D. Pedoe, *A course of Geometry for Colleges & Universities*, Cambridge University Press, London, 1970.
100. G. Salmon, *A Treatise on Conic Sections*, Chelsea, N.Y., 1954.
101. P. F. Smith and A. S. Gale, *The Elements of Analytic Geometry*, Ginn, Boston, 1904.

102. N. Vasilev and V. Gutenmacher, *Straight Lines and Curves*, Mir, Moscow, 1980.
103. W. A. Wilson and J. I. Tracey, *Analytic Geometry*, Heath, Boston, 1937.

Geometry (Solid):
104. R. J. T. Bell, *An Elementary Treatise on Coordinate Geometry of Three Dimensions*, Macmillan, London, 1912.
105. P. M. Cohn, *Solid Geometry*, Routledge and Paul, London, 1965.
106. N. Altshiller-Court, *Modern Pure Solid Geometry*, Chelsea, N.Y., 1964.
107. A. Dresden, *Solid Analytical Geometry and Determinants*, Wiley, N. Y., 1930.
108. W. F. Kern and J. R. Bland, *Solid Mensuration with Proofs*, Wiley, N.Y., 1938.
109. L. Lines, *Solid Geometry*, Dover, N.Y., 1965.
110. G. Salmon, *A Treatise on the Analytic Geometry of Three Dimensions*, Chelsea, N.Y., 1954.
111. D. T. Sigley, W. T. Stratton, *Solid Geometry*, Dryden, N.Y., 1956.
112. C. Smith, *An Elementary Treatise on Solid Geometry*, Macmillan, London, 1895.

Graph Theory:
113. B. Andrasfai, *Introductory Graph Theory*, Pergamon, N.Y., 1977.
114. B. Bollobas, *Graph Theory, An Introductory Course*, Springer-Verlag, N.Y., 1979.
115. J. A. Bondy and U. S. R. Murty, *Graph Theory with Applications*, American Elsevier, N.Y., 1976.
116. F. Harary, *Graph Theory*, Addison-Wesley, Reading, 1969.
117. R. Trudeau, *Dots and Lines*, Kent State University Press, Ohio, 1976.

Inequalities:
118. O. Bottema et al, *Geometric Inequalities*, Wolters-Noordhoff, Groningen, 1969.
119. G. H. Hardy, J. E. Littlewood, and G. Pólya, *Inequalities*, Cambridge University Press, Cambridge, 1934.
120. D. S. Mitrinovic, *Analytic Inequalities*, Springer-Verlag, Heidelberg, 1970.
121. D. S. Mitrinovic, *Elementary Inequalities*, Noordhoff, Groningen, 1964.

Theory of Numbers:

122. G. E. Andrews, *Number Theory*, W. B. Saunders, Philadelphia, 1971.
123. R. D. Carmichael, *The Theory of Numbers and Diophantine Equations*, Dover, N.Y., 1959.
124. L. E. Dickson, *History of the Theory of Numbers I, II, III*, Stechert, N.Y., 1934.
125. G. H. Hardy and E. M. Wright, *Introduction to the Theory of Numbers*, Clarendon Press, Cambridge, 1954.
126. I. Niven and H. S. Zuckerman, *An Introduction to the Theory of Numbers*, Wiley, N.Y., 1960.
127. H. Rademacher, *Lectures on Elementary Number Theory*, Blaisdell, N.Y., 1964.
128. W. Sierpinski, *Elementary Theory of Numbers*, Hafner, N.Y., 1964.
129. J. V. Uspensky and M. A. Heaslet, *Elementary Number Theory*, McGraw-Hill, N.Y., 1939.

Trigonometry:

130. H. S. Carslaw, *Plane Trigonometry*, Macmillan, London, 1948.
131. C. V. Durell and A. Robson, *Advanced Trigonometry*, Bell, London, 1953.
132. E. W. Hobson, *A Treatise on Plane and Advanced Trigonometry*, Dover, N.Y., 1957.
133. T. M. MacRobert, W. Arthur, *Trigonometry; I (Intermediate), II (Higher), III (Advanced), IV (Spherical)*, Methuen, London, 1938.

Other:

134. R. W. Benson, *Euclidean Geometry and Convexity*, McGraw-Hill, N.Y., 1966.
135. E. R. Berlekamp, J. H. Conway, and R. K. Guy, *Winning Ways*, Vols. I, II, Academic Press, London, 1982.
136. M. P. Gaffney and L. A. Steen, *Annotated Bibliography of Expository Writing in the Mathematical Sciences*, M.A.A., Wash., D.C., 1976.
137. A. Gardner, *Infinite Processes-Background to Analysis*, Springer-Verlag, N.Y., 1982.
138. I. M. Gelfand, E. G. Glagoleva, and E. E. Shnol, *Functions and Graphs*, M.I.T. Press, Cambridge, 1969.
139. S. I. Gelfand et al, *Sequences, Combinations, Limits*, M.I.T. Press, Cambridge, 1969.
140. W. Gellert et al (eds.), *The VNR Concise Encyclopedia of Mathematics*, Van Nostrand Reinhold, N.Y., 1977.
141. S. Goldberg, *Introduction to Difference Equations*, Wiley, N.Y., 1958.

142. L. A. Lyusternick, *Convex Figures and Polyhedra*, Dover, N.Y., 1963.
143. I. Niven, *Maxima and Minima Without Calculus*, The Dolciani Mathematical Expositions, Vol. VI, M.A.A., Wash., D.C., 1981.
144. T. L. Saaty, *Optimization in Integers and Related Extremal Problems*, McGraw-Hill, N.Y., 1970.
145. W. L. Schaaf, *Bibliography of Recreational Mathematics*, Vols. I, II, III, IV, N.C.T.M., Wash., D.C., 1965, 1972, 1973, 1978.
146. S. Schuster, *Elementary Vector Geometry*, Wiley, N.Y., 1962.
147. A. Soifer, *Mathematics as Problem Solving*, Ctr. for Excellence in Math. Ed., Colorado Springs, 1987.

Also:

The entire collection of the New Mathematical Library (now 33 volumes —see page vi of this book), available from The Mathematical Association of America, is highly recommended.